Ecologism

Ecologism

An Introduction

Brian Baxter

Georgetown University Press
Washington, D.C.

To my mother, Dorothy,
and to the memory of my father, James

Georgetown University Press
Washington, D.C. 20007

First published in the United Kingdom
by Edinburgh University Press

Typeset in New Baskerville
by Norman Tilley Graphics, Northampton
and printed and bound in Great Britain
by MPG Books, Bodmin

A CIP record for this book is available
from the Library of Congress

ISBN 0-87840-781-2 (paperback)

Contents

Acknowledgements vii

1 Introduction 1

Part One – Theory
2 Metaphysics 15
3 Biology and Ecologism: The Case of Sociobiology 34

Part Two – Morality
4 The Moral Theory of Ecologism 51
5 Moral Considerability and Moral Trade-offs 74

Part Three – Politics
6 Ecologism's Political Philosophy: Human Nature, the Human Predicament and Political Morality 103
7 Ecologism's Political Philosophy: Political Morality and Meta-issues 120
8 Ecologism and Contemporary Political Philosophy: Utilitarianism, Rawlsian Liberalism and Libertarianism 145
9 Ecologism and Contemporary Political Philosophy: Marxism, Communitarianism and Feminism 163

Part Four – Political Economy
10 Can Ecologism Transform Capitalism?: Sustainable Development, Ecological Modernisation and Economic Democracy 187

11 Alternatives to 'Greened' Capitalism: Market Socialism and Global Ecology 210
12 Conclusion 231

Bibliography 236
Index 243

Acknowledgements

AT A RECENT Academy Awards ceremony the Hollywood film star Kim Basinger said in her acceptance speech that she just wanted to thank everbody she had ever met in her entire life. The author of a first book will recognise the sentiment behind this. In my case I have to thank my parents for the love and encouragement they gave me throughout my childhood and youth, especially for setting me an example of their enjoyment of the natural world. This enjoyment, in which I quickly came to participate, is something which has finally borne fruit in my professional career in the shape of this book and my ongoing interest in the problems of the environment.

I wish to express my gratitude to those who taught me philosophy at the University of Oxford – John Simopoulos, Geoffrey Warnock, Richard Malpas, Stuart Hampshire and my D.Phil. supervisor, James Urmson – for helping to inculcate in me a life-long fascination with philosophical issues. The first draft of this book was written during a period of sabbatical leave in 1997, so thanks must go to my Department and its then Head, Tony Black, for allowing me this crucial time away from teaching and administration.

My special thanks go to John Barry who read and commented on the entire typescript, making many invaluable comments and suggestions throughout and providing a very detailed response to Chapter 8, where his comments forced me to be much more explicit on the issues of social justice, ecological justice and the connections between them. Our views on the matters discussed in this book have independently been developing along parallel paths and it has been a pleasure to experience his enthusiasm for, and knowledge of, the subject. Of course, responsibility for the content of the book remains entirely mine.

Nicola Carr of Edinburgh University Press offered the first encouragement to me to produce a book in this area, and has proven to be a pleasure to work with. Susan Malloch has provided essential secretarial help at crucial points. Thanks to our friends Colin and Ursula Doherty for the gift of a cafetière whose contents helped to fuel the creative processes during writing.

Finally, the deepest of thanks to my wife, Lynn, for providing the continuous love and encouragement which have enabled my thoughts to reach expression on the printed page.

1

Introduction

ECOLOGISM IS A NEW STAR in the ideological firmament. I first encountered the term in Andrew Dobson's *Green Political Thought.*[1] He there argued that the position to which he gave the name was sufficiently comprehensive and systematic to count as an ideology and was distinct from other ideologies in crucial respects. It embodies ideas and prescriptions which have been formulated over the last thirty or forty years by environmentally concerned thinkers of all kinds drawn from the humanities and social and natural sciences.

This book is written in the conviction that Dobson is correct to suggest that a new ideology has crystallised out of the concerns many thoughtful people have been articulating about how human beings are related, and are relating, to the planet which is, at least for the foreseeable future, their sole home. Of course, as anyone with knowledge of contemporary environmental concerns is aware, what has brought environmental matters to the fore is a developing sense among many people of at least crisis, and possibly impending catastrophe, on a number of environmental fronts. There is a litany of problems which it is by now customary to rehearse in the introduction to books on environmental matters, whether the author is concerned to show that they involve genuine crises and to explain how they may be dealt with, or is concerned rather to show that they are expressions of exaggerated worries, or perhaps even non-existent.

Instead of running through this list one more time, I will mention the concern which actuates my own environmental interests, and which attracted me to the possibility of articulating a worked-out ideological position to deal with it. This is the fear that human beings are in the course of perpetrating a 'sixth extinction' on this planet,[2] to rival those occasionally produced by natural processes aeons before the human species appeared.[3]

The problem to which this fear points is sometimes referred to, in somewhat clinical terms, as the problem of 'preserving biodiversity'. It appears to be becoming acute in the face of the rapidity of human population growth and that intensification of world economic activity with which we have become familiar since World War Two. To many it now seems that we have inherited a planet teeming with life-forms of immense variety, life-forms to which we are intimately connected by evolutionary history, whose continued existence we have taken too much for granted and which appears now in many cases to be under serious threat.

One may, of course, reject the factual claims embodied in this view as a 'scare story'. But if you believe that there is a real process under way which threatens large-scale extinctions, as I do, there are still various possible responses one may make to it. One is to argue that even if mass extinctions are taking place, there is no overall threat to life in general or to major biological functions on this planet. It may be said that the biosphere of the planet has coped with worse in the past, and human life in particular is under no real threat from the disappearance of even large numbers of other species caused by humanity's onward march to a better life.

A second response is to accept that there is a crisis and to claim that the future well-being of human beings is threatened by it. This may be supported on the basis of the argument that it threatens to undermine at least some vital life-support functions of the planet, ones which we will find it difficult, if not impossible, to repair or replace. A variety of this second response is to go on to emphasise that more than just our physical well-being is under threat here. Our cultural and spiritual lives are thoroughly entwined, it may be argued, with the living tissue of the planet. If that is severely damaged, our psyches as well as our bodies may suffer irreparable harm.

These two responses look at the matter solely from the point of view of how human beings may suffer from a severe depletion of the planet's stock of life-forms. The best of these human-being-oriented, or 'anthropocentric', responses is the variety of the second response mentioned in the previous paragraph. However, not even this is wholly adequate to capture my own view of what is wrong in this 'extinction' crisis.

This view is not unique to me, but has been articulated by many others in a variety of guises and for various reasons. It is the sense that the world of non-human nature has value in itself, irrespective of its contributions to (or, for that matter, detractions from) human well-being. The harm done by human extinctions is harm done to the species, and its individual members, rendered extinct. If there is a moral wrong done here it is done to those creatures, as well as to other creatures, including human beings, whose interests are severely damaged by the knock-on effects of the disappearance of the life-forms in question.

This, at least, is the position to be defended in this book. It began life as an attempt to relate the issue of the preservation of biodiversity to the subject-matter of political philosophy, to which it has only recently been connected in anything like a concerted way. In the course of attempting to work out my ideas on this I have found that the whole range of metaphysical, moral, political, economic and cultural issues with which political philosophy has to deal has to be considered. Hence this book has turned into the elucidation and defence of the new political ideology of ecologism.

This interconnectedness of issues is no cause for surprise to those concerned with the examination of ideological constructs. It is even less of a surprise to the defenders of ecologism who, taking their cue from the science of ecology, are the first to emphasise that the reality for life-forms within a single biosphere is that of interconnectedness, one of the major themes of ecologism as an ideology.

It should be said at this point, however, that not all environmentally concerned thinkers believe that the development of an environmental ideology is desirable. For example, Bob Pepperman Taylor has recently argued against the ideological orientation of many of those concerned to develop environmental ethics, specifically those offering a 'biocentric' ethic, which seeks to displace the human species from the centre of moral concern and replace it with the Earth's biosphere.[4] He has specific criticisms to make of the contents of their view, some of which are certainly sound. But more challengingly, from the point of view of this book, he argues that the development of policies to deal with environmental problems should derive from the practice of what he calls 'pragmatic democratic politics', rather than from the development of a philosophical theory by academics and others.[5]

The main charges he lays against the development of ideology as such are idealism and exclusivity. That is, those who believe that an ideology is necessary in order to change people's world-view, and thus their behaviour, are wedded to a hopelessly inadequate understanding of the factors which produce human behaviour. He does not explain what precisely these factors are, but he is clear that just changing a world-view, even if that could be done, may leave what people actually do completely unchanged. Secondly, ideology is inherently exclusive. It divides people into believers and non-believers, with the former inevitably looking down on the latter as benighted. It also divides believers into purists and realists, with the former accusing the latter of treason, and the latter accusing the former of naivety. Ideologists seek to convert their opponents wholesale, by force in the last resort. Together, these make ideological politics inherently anti-democratic and lead ideologists into the sterile backwaters of politics.

Pragmatic democratic politics, however, aims to include people, not exclude them. Thus, Taylor says, 'Democratic politics insists that common ground be sought, that solutions to political problems be found that as many people as possible can accept, however grudgingly.'[6] However, Taylor's diagnosis looks to be misplaced. What he is taking issue with is a certain kind of ideological position, of which there are many varieties, which sets no particular store by democratic debate and persuasion. Ideology as such involves no necessary commitment to such a view. Indeed the placing of democratic debate and persuasion at the heart of politics is central to the ideology of liberalism. As we will be seeing later on, when we turn to consider the view which ecologism is committed to take with respect to democracy, this ideology also is committed to democratic debate.

But also, pragmatic democratic politics is itself only possible if the participants hold views, perhaps strongly held, about values, problems, acceptable solutions and so forth. There is no reason to exclude ideological positions from this. Otherwise Taylor's position is in danger of concluding that only people with no particular values or beliefs about anything can participate successfully in such politics. Of course, there are the dangers of exclusivity to which he points. But they are emphatically not a reason for avoiding all contact with ideology.

Finally, the charge of a hopelessly naive idealism is less telling than he suggests. The view that changing people's ideas can change their behaviour is not as unsustainable as he supposes. For one thing, idealism as a theory of social change is presumably to be contrasted with materialism, which offers the view that social changes (and changes in consciousness) are brought about by changes in the system of material production, or the economy, for short. But an examination of how economies function shows how crucial is the production of ideas to the success of economic activity. Particularly in a market economy, enormous effort goes into changing people's values, beliefs, desires and so on via advertising, manipulation of the media and other channels of ideational formation. Thus, at the very least, it is not prima facie obvious that an ideology, which certainly addresses the ideas which people hold, is simply beside the point in producing social change. It is certainly plausible to suggest that changes in ideas are not a sufficient condition of changes in people's behaviour. But it seems equally plausible to suggest that they are a necessary condition. This book, at any rate, is written on that premise.[7]

However, it offers only one version of ecologism. There are others which differ in important respects from what is set out in the following pages. To begin with, although Dobson, in his characterisation of the ideology, cited the moral claim just mentioned as central to it, he gave equal billing to the claim that the planet is in important respects finite, so that endless expansion of human population and material production is held to be impossible. Thus a key claim of ecologism, on Dobson's presentation of it, is held to be the need to live within those limits.

This claim is not absent from the account given in this book. But it is in a decidedly subordinate position, mainly because it involves a whole series of further empirical claims whose truth is still open to question by reasonable people, at least when those people are considering solely human material well-being. For this reason, I believe that ecologism would be well advised not to give the limits theses too much prominence. If they are true, then of course they give strong reasons for rethinking the way we live now in fundamental ways. But they may not be true, and if they are refuted then there will be the temptation to believe that the case for ecologism is also thereby refuted.

But it is not. At least, the version of ecologism presented here can weather such a setback, because that version amounts primarily to the case for a drastic reconsideration of something non-empirical, namely the way we think about moral matters. Ecologism is here presented as fundamentally a thesis about the moral considerability of other living creatures. Hence, even if, contra the limits theses, we discover endless cheap energy supplies, invent new cornucopian techniques, such as nanotechnology,[8] find ingenious ways to dispose of our (minimised) wastes, discover ways to extend our physical presence under the earth, in the sky, in space, under the waves and so on, so that our living-space can be endlessly expanded, still the fundamental moral claims of non-human life-forms will not be affected. Such claims, being moral, cannot be ignored or deferred. Hence, the urgency imparted to environmental theories by the limits theses is not lost, but given an alternative basis. The nature of the moral claims is also such that they impose demanding moral constraints upon what we are permitted to do.

Moral constraints are thus stronger than physical ones. But they are in many ways harder to establish. In the area of moral philosophy there is a dearth of knock-down arguments. Convincing moral positions are rather achieved by a protracted process of argumentation and reflection. They achieve their place in the pantheon of moral theories by containing promising lines of argument as well as embodying claims which carry weight with people when they turn to consider them.

Arguably the moral claim inherent in ecologism has attained that position. At the very least, devotees of rival moral positions have now to have some argument to weaken or refute the claim of the moral considerability of the non-human. They cannot simply ignore it and still hope to have a position which possesses any show of adequacy. However, there are no knock-down arguments in this book either. Difficulties and dilemmas remain in the moral position of ecologism, as they do in every other moral position.

A third strand in the ideology of ecologism, in addition to the moral and 'limits' theses just mentioned, is the claim, already noted in passing, of human interconnectedness with the biosphere of this planet. Some environmental philosophies, such as that which is to be found in the work of Bryan Norton, give the greatest emphasis to this strand.[9] On this view, it is our failure to grasp that we have not severed these connections, and arguably cannot, at least while we remain on this planet, which is the prime cause of our environmental predicament.

Such a claim has various components, for our interconnections with the living world around us are not just physical but also cultural and spiritual. Once we grasp this, and develop a properly contextual view of ourselves, then we may begin to employ our scientific and technological ingenuity, which we still need to depend on for the improvement of human life, in a way which does not threaten the integrity of the biosphere. This view lends support to the claim that what we need most is an enlightened anthropocentrism, which sees where our real needs lie. The claim concerning the moral considerability of the non-human is either not accepted, or is held at arm's length. This version of ecologism is discussed more fully in Chapter 4.

The version of ecologism presented in this book accepts the 'interconnectedness' theses, but does not allow them to displace the moral claim from central position. The fact of our interconnectedness with the rest of nature is of importance to ecologism because it suggests how environmental destruction adversely affects human and non-human well-being and thus marks out the area of our moral responsibility. But part of what is meant by attributing moral considerability to the non-human is that we ought to conduct ourselves in certain respects towards the latter even if we are not interconnected with it, or do succeed in severing our present links with it.

These ideological variations within a common family are, as Tim Hayward has argued, always to be expected.[10] Every ideological position will exist in various versions, depending upon which features are kept and which rejected, which emphasised and which downplayed. This is no less true of ecologism in particular and the wider family of what one might call 'Green political theory' in general.

Thus, as we have just noted, there are versions which are based on 'enlightened' anthropocentrism, which emphasise the interconnection of human and non-human well-being and which are very chary of attributing any very full-blooded sense of moral worth to the non-human. The versions expressed by Hayward himself, and by ecosocialists in general,[11] are in this wing of the Green ideological mansion. It is the version which may be expected to have the most immediate appeal, since its main emphasis is on interconnectedness, especially in the biological arena. This version thus appears to be both scientifically well attested and to require for its acceptance only enlightened self-interest and a sensitivity to the non-human world of nature, something which many people appear to possess, rather than any extensive rethinking of moral positions.

Religion-based stewardship positions are another version of Green political theory with some immediate appeal to large numbers of people, relying as they do upon historically validated interpretations of religious beliefs, and again requiring an extension, rather than a rethinking, of fundamental moral positions.[12]

Hence, the form of ecologism offered here has rivals within green political theory in addition to the rivals outside it. It puts prime emphasis on the moral claim; strongly supports the interconnections theses; and is well disposed towards the 'limits' theses, although it remains still open-minded towards them. It is also science-oriented and naturalistic. It sets great store by the findings of natural science for the light these throw on the moral standing of the non-human, the ways in which human beings are a part of nature and the ways in which human activity may be damaging the environment. Its naturalism – the attempt to understand and explain human values as produced by natural, biological, processes – leads it to reject dualism between natural scientific and social scientific approaches to the study of human beings.

For this reason, it adopts an attitude towards human reason akin to that put forward by Hayward.[13] That is, it emphasises the need for reason, not its abandonment, in our conduct *vis-à-vis* the non-human world. But this is reason employed in the light of the moral claims of that world, our interconnection with it and the strong possibility of there being limits to our making it over entirely in the light of purely instrumental considerations.

The three theses which we have attributed to ecologism all point towards the nefarious implications of three forms of human arrogance, namely the beliefs that: the non-human is of no moral account; we can eventually do without the non-human natural world; there are no limits to what human activity can achieve. But these forms of arrogance, which ecologism is certainly avid to combat, are not the essential offspring of reason. Indeed, it takes the employment of reason, especially in the scientific sphere, to see precisely where the arrogance lies in these views.

One of the aims of ecologism, science-oriented as it is, is to rehabilitate a specific view of science, which Norton has eloquently delineated in his discussion of the scientific views of John Muir.[14] This sees science not as a handmaiden of purely instrumental, technological thinking, but as a mode of wonder, revealing the glory of the world. Hence the view of ecologism presented here does not seek the overthrow of the Enlightenment, especially not the role of critical reason. It seeks rather its reorientation in a more morally and scientifically adequate direction.

One may sum up the brief characterisation given so far of the view of ecologism presented in this book by employing a distinction recently developed by John Dryzek.[15] He distinguishes, within green radical discourse (of which this is intended to be a specimen) between 'Green romanticism' and 'Green rationalism'. Green romanticism aims to save the environment by bringing about changes in the attitudes, values and beliefs of human individuals, so that they begin to relate differently than hitherto to their natural context. Green rationalism, he says

> may be defined in terms of its selective and ecologically guided radicalisation of Enlightenment values ... Rationality is also a matter of open-ended and critical questioning of values, principles and ways of life – which opens the door to critical ecological questioning.[16]

From this stems Green rationalism's commitment to seeking ways to alter societies' social, political and economic institutions in the direction of greater ecological adequacy.

The version of ecologism given here lies within the Green rationalist ambit, although it differs from some others within that ambit in the precise way in which it understands and locates the three strands mentioned above. It also has to be said that it occasionally employs arguments, such as the discussion of the moral importance of wonderfulness in Chapter 4, which might more readily be labelled 'Green romantic'. Dryzek notes that Green rationalism is still an emerging discourse. The arguments of this book may be understood as a contribution to that emergence.[17]

It may be helpful at this stage to give a summary of the main points of the position developed below. It is one which:

(1) accords moral considerability to non-human beings, but attributes it to them to different degrees;
(2) accords the highest degree to human beings, but requires human beings to take moral account of the non-human;
(3) makes human well-being of central concern, but insists upon understanding that well-being in contextual terms;
(4) views human beings as highly interconnected with the non-human, physically, culturally and spiritually, but rejects the view that the moral standing of the non-human stems from this fact;
(5) argues that the moral considerability of the non-human requires a new political philosophy in which issues of justice between human and non-human are addressed (in Chapters 8 and 9

ecologism is brought into direct critical contact with 'mainstream' political philosophy);

(6) suggests ways in which political structures and other social practices, especially the economic, require extensive modification to attain this;

(7) emphasises moral considerations, rather than resting a great deal of weight upon prognostications of ecological crisis, although the latter are not rejected;

(8) does not seek to make major claims about limits to human activities, although what is said is compatible with there being such limits.

The book also addresses some issues which have not been quite so much in the forefront of environmental debates as they might have been. In particular, it seeks to explore the implications of the view that human beings are part of nature, not set apart from it, and to explore the implications of this idea for human self-understanding and well-being. In particular, it assesses whether the most dramatic of the naturalistic positions, sociobiology, is one which ecologism should be willing to countenance, or whether this naturalistic position at least should, from the point of view of ecologism, be struck off the list of theoretical possibilities for self-understanding.

The table of contents indicates the structure of the book, which sets out to offer a coherent and connected account of the ideology of ecologism. Inevitably, some themes are only sketchily indicated. Many would require books to themselves. However, the aim is to ascertain in what theoretical and practical directions ecologism should be looking for solutions to what are immensely complex problems. An ideology is a map to help guide one through an infinitely particular world. Which map to use depends on the purpose of one's journey. This map aims to give a view of the main features of the terrain and their relationship to each other, for the purposes of general orientation.

One final terminological point. The ideology is called 'ecologism', which usefully suggests its connection with the scientific discipline of ecology while indicating its comparability with other ideological '-isms'. However, its supporters cannot properly be called 'ecologists', since that term is already used to refer to the practitioners of the scientific discipline. They are sometimes called 'political ecologists', which suggests that the ideology should be called 'political ecology', a term occasionally used. This makes sense, but runs the risk of confusing supporters of the ideology with scientific ecologists who happen also to be politically minded, and who may as individuals be strongly opposed to this new ideology.

Throughout this book, therefore, in order to reduce the possibility of confusion, I nearly always use the name of the ideology (as in 'ecologism supports the view ...') rather than refer to what its supporters, 'political ecologists', believe.

Notes

1. See the Introduction to Andrew Dobson, *Green Political Thought*, 2nd edn (1995). 'Ecologism' is such a new term that its use has not settled down yet. For example, Arne Naess uses it pejoratively in his *Ecology, Community and Lifestyle* (1989) to refer to 'excessive generalisation of ecological concepts and theories' (p. 39).
2. The allusion here is to Richard Leakey and Roger Lewin, *The Sixth Extinction* (1996).
3. A good general discussion of extinctions may be found in David Raup, *Extinction: Bad Genes or Bad Luck?* (1991) and in E. O. Wilson, *The Diversity of Life* (1992). In addition to Leakey and Lewin 1996, a general assessment of the present situation is to be found in John Lawton and Robert May (eds), *Extinction Rates* (1995). Gill Aitken, *Extinction* (1996) provides a very useful philosophical analysis of what is different about the present extinction episode.
4. Bob Pepperman Taylor, 'Democracy and environmental ethics' (1996).
5. Taylor 1996, p. 101.
6. Taylor 1996, p. 101.
7. Taylor himself appears ambivalent on the matter. He castigates biocentrists for not attempting to promulgate their moral views, and compares them unfavourably with Romantic critics of capitalism in that respect. He may well have a point, but it is hard to see how he can berate them thus and simultaneoulsy castigate the whole notion of 'changing ideas'.
8. Nanotechnology is the creation of materials by directly manipulating atoms.
9. Bryan Norton, *Toward Unity among Environmentalists* (1991).
10. Tim Hayward, 'What is Green political theory?'(1996).
11. See, for example, David Pepper, *Eco-Socialism* (1993). Robyn Eckersley, *Environmentalism and Political Theory* (1992), has an excellent discussion of socialist thought and environmental concerns.
12. Religious stewardship positions are to be encountered in Herman Daly

and John Cobb, *For the Common Good* (1990) and in Robin Attfield, *The Ethics of Environmental Concern* (1991).

13. See Tim Hayward, *Ecological Thought* (1995).
14. Norton 1991, pp. 32–5.
15. John S. Dryzek, *The Politics of the Earth* (1997), p. 172.
16. Dryzek 1997, p. 172.
17. Dryzek 1997, p. 173.

Part One

THEORY

2

Metaphysics

THIS CHAPTER AND THE NEXT will be consciously exploratory. I will reach some conclusions about the relations of ecologism to metaphysics and natural science, but much of what we can conclude from the analysis will necessarily be provisional. My aim will be to consider some options for the metaphysical and scientific foundations of ecologism, suggesting, where the argument permits, which of the options are ones to which ecologism should aim to be hospitable, and which are ones it would do well to avoid.

The tentative and exploratory nature of these chapters will not be a serious drawback. As indicated in the introduction, ecologism is primarily a moral doctrine, albeit one informed by certain scientific and metaphysical tendencies. However, I argue that ecologism can afford to be flexible on these matters, even though an ideal theory would move seamlessly from metaphysical doctrine to scientific theorising, moral and political prescriptions and a total view of the world in which humanity's place and the meaning of human life are both set forth clearly. We do not yet have that kind of ideal theory, and it is not too difficult to show that other doctrines which may claim to exemplify it in fact do not – tensions and contradictions abound both within and between the areas of theory just mentioned. In this respect, then, ecologism is at least no worse off than are other influential political philosophies.

Let us now turn to consider the first of these issues, which is whether any metaphysical theory is available to ecologism which can provide the underpinning for its whole range of concerns. One excellent example of metaphysical theorising which aims to provide an 'ideal' theory of the type just canvassed is Freya Mathews' splendid book *The Ecological Self*.[1] As will shortly become apparent, I am unconvinced by the main claims of this book and by similar claims of other theoretical perspectives, such as some of the 'deep ecology' views associated with such philosophers as Arne Naess.[2] However, Mathews' book is highly suggestive and poses the key issues in a way which anyone working in this area will do well to ponder.

It is, first of all, crystal clear that ecologism is committed to the rejection of at least one metaphysical position, namely atomism. As Mathews conclusively demonstrates, atomism, whether as a metaphysical theory about the ultimate form of reality, or in its scientific manifestation as Newtonian physics and as individualistic social science, has unacceptable implications as far as ecologism is concerned. Without going into the details of these arguments, it can be said that the problem with atomism is that it presents a picture of the human and non-human world which fails to encompass the interconnectedness of things. In particular, it encourages human beings to view themselves as in some important way disconnected from the natural world around them, with disastrous results for both the natural world and human beings themselves.[3]

This raises the constant ecologistic motif of holism, in which parts relate to wholes in a manner which is logically mutually determining. That is to say, on the one hand the whole determines the nature of its parts, so that no accurate description of the parts can be given without reference to the wholes of which they form a part. The parts 'function towards the whole', to quote J. C. Smuts, and this function is a necessary part of their description.[4]

The whole, in turn, possesses an overall character which is determined by the nature of the parts, but not their nature considered in isolation from the whole. Hence the whole is not the mere sum of its parts considered in such isolation. The character of the whole cannot be analysed simply into the totality of separate effects of the parts, for this ignores their functional role just referred to.

By contrast, in atomistic analyses, wholes are fully determined by the summation of the separate effects of their parts. The parts can thus be characterised in isolation from the wholes which they constitute, and the wholes are constituted in a purely additive manner from those parts.

As Mathews explains, the holistic view primarily analyses reality in terms of systems.[5] Systems theory is available to analyse systems in terms of various processes designed to maintain them in existence in the face of potentially disruptive effects impinging upon them from the outside. The theory of the biological organism is one example of how this works. Organisms are physical systems operating on the basis of various feedback mechanisms to keep themselves in existence by maintaining certain key internal states within a specific range – temperature, chemical composition, and so on. The general name for such processes is homeostasis – 'the same state'.

Some systems – such as machines – contain feedback devices, such as thermostats, to produce homeostasis. But these systems are not self-maintaining.[6] That is, they have to be operated by external forces – human beings usually – to bring the feedback mechanisms into play. When an automobile runs out of petrol, to use a stock example, it does not actively seek a source of fuel itself, a human driver has to do this for it. By contrast, an animal does actively seek food when it is hungry.

Systems can be located in an enormous variety of situations. Holistic approaches can be found to the study of physical, biological and social phenomena. Systems theory, as a holistic approach, grants reality to the systems it studies. Since they are not just the sums of their parts, they have a real existence over and above their component materials. They have their own histories and can enter into a wide range of causal relationships with other systems, and with things which are not systems.

In some fields the existence of systems has been hotly disputed by theorists who are wedded to atomistic approaches to the segment of reality in question. Mathews analyses the effect of atomistic approaches to the study of human societies, for example. Classical liberalism, as exemplified in the work of such philosophers as Locke, appears to be committed to a non-holistic view of human societies. The latter are taken to be simply agglomerations of human individuals, construed in atomistic terms as analysable in abstraction from the societies of which they are members. Hence, on this view, to refer to society is simply a short-hand way of referring to the individuals which make them up.[7]

In general, as socialists have often insisted, liberals, with their commitment to an atomistic individualism, have tended to be blind to the existence of systems in the social sphere, and thus to the causal effects of such systems on individuals. Liberals, in turn, have frequently lampooned the non-liberal tendency to assign blame for what liberals see as individual human failures to the malign effects of some system, such as the economic system, the system of class oppression, the system of patriarchy and so on. Perhaps holism does inculcate a tendency to find systems where there really are none. However, ecologism is strongly orientated to a systems approach and indeed to insist that its philosophical rivals, even the sytems-oriented socialists, have failed to note either the existence or the importance of certain key systems – most notably ecosystems in general and the Earth's biosphere in particular.

It is this orientation which Mathews seeks to vindicate by demonstrating that a holistic metaphysics can be consistently developed which locates systemic features from the level of the cosmos as a whole down to the level of the simplest organisms. Further, these systems are the dynamic kind mentioned above. That is, unlike machines, they possess the capacity to be self-realising, actively seeking to secure their own existence and achieve their own flourishing. The cosmos, from top to bottom, is thus not a piece of giant clockwork, requiring an external force to wind it up. Matthews employs the term 'selves' to refer to such self-realising systems, whether or not they possess the consciousness or self-awareness present in organisms such as human beings and other complex mammals.[8]

In seeking to establish this position, Mathews claims the support of modern scientific cosmology, based on the ideas to be found in relativity theory and quantum mechanics. That is, she convincingly shows that the most prestigious form of natural science has long ago abandoned atomism. The Einsteinian world is, on her analysis, a monistic and holistic one. There is a single cosmic reality – spacetime – which is unfolding in accordance with its own inner nature so as to produce self-realisation via self-diversification into particular entities. Specifically, the particular physical entities with which we are familiar, from galaxies to quarks, are created out of spacetime by a process of geometrical infolding, in the manner suggested by the analysis of gravity in Einstein's general theory of relativity. Such infoldings – regions in which spacetime has become highly curved – are related to spacetime in the holistic manner outlined above. Their geometrical nature necessarily links them to the spacetime of which they are subregions. Their interconnectedness with each other and with the cosmic spacetime of which they are part is thus established at the most fundamental level of physical reality.[9]

The scientific theory here referred to – geometrodynamics, to use the term for it employed by Mathews – is then related to the metaphysics via the theory of monism to be found in the work of Spinoza.[10] Mathews characterises the cosmic spacetime of modern physics as the sole entity which possesses the status of substance-hood.[11] A substance is conceived of as a bearer of attributes. Unless we posit such a phenomenon, we will have to suppose that attributes might be able to exist without being the attributes 'of' anything. This supposition, Mathews argues, lands us in the predicament of being unable to explain the difference between a purely abstract universe and a really existing one.[12] Hence we need at least one substance. However, we can now see from the cosmology of modern physics that we do not need more than one. All the particular things which exist are not substances in their own right, as atomism was led to suppose, but attributes of cosmic spacetime, related holistically to that whole.

This rather austere picture which focuses on the logical category of substance is enriched and taken into the moral and spiritual dimension by the claim that the cosmic spacetime is a dynamic system which qualifies as a 'self' – it is self-maintaining and self-realising.[13] Its self-maintaining nature means that no external creator is called for. Its self-realising nature means that it actively produces its internal differentiation into particular entities – many of which are also 'selves'. At this point we discover a metaphysical reason for setting store by biodiversity within the Earth's biosphere (itself analysable as a self-sustaining self, as suggested by Lovelock's 'Gaia' hypothesis).[14] For the self-realisation of the cosmic self is matter of its internal differentiation into as many particular forms as possible.

Value emerges at the point where we encounter selves. For, Mathews argues, a self necessarily values its own existence, as evidenced by its necessary commitment to self-preservation. This existence is valued by the self for its own sake. Another word for this is 'intrinsic' value. Hence, selves necessarily possess intrinsic value. It then analytically follows from the meaning of 'intrinsic value' that such entities are, *ceteris paribus*, to be preserved by creatures capable of recognising this value – namely moral creatures, such as our-selves.[15] This is an important part of Mathews' argument and we will return to examine it in Chapter 4.

The conclusion to be drawn from this analysis for human beings is, of course, that the most accurate picture of the universe drawn from the physical and biological sciences, and located firmly within a monistic metaphysic, is one which depicts us as interconnected with the rest of the universe at all levels of analysis. Our existence and flourishing cannot, therefore, be separated conceptually or morally from the existence and flourishing of the rest of the universe. Specifically, on this planet, our existence and well-being are highly interconnected with those of other organisms, species and the entities analysed by the science of ecology – ecosystems and the biosphere as a whole. Both morality and prudence counsel us to adopt a caring, preservative attitude towards the world of selves with which we are connected, however lowly some of those selves appear to be.

Finally, Mathews draws from her theory conclusions about the meaning of human life which can be inferred from our interconnection with other self-realising selves, including the cosmic self. Our participation in the self-realisation of a cosmic self is enough to endow our lives with the kind of meaning which, in the spiritual and religious aspects of our lives, we have long been seeking.[16]

As will be apparent, this theory is a bold and comprehensive attempt to produce the 'ideal' theory mentioned above. Its appeal to contemporary science is a notable feature of its approach. As Mathews notes, while many of the conclusions she reaches may be endorsed by other traditions of thought, especially by some of the ancient eastern religions, for most people living under liberal capitalism such conclusions need to be drawn from the crucial elements in our own culture, than which there is none more central or prestigious than natural science.[17]

Although she presents a compelling amalgam of modern physics and Spinozist metaphysics, there are problems lurking everywhere in this project. First, the metaphysics of monism, like all metaphysics, is esoteric, and relies on principles and arguments whose validity is hard to determine conclusively. It is fair to say that this metaphysic is far from central to our cultural traditions, and thus its acceptance will involve both understanding of some abstruse theorising and a new intellectual departure for most people in our culture, whose 'common sense' metaphysic probably amalgamates elements of atomism and parts of the Judeo-Christian world-view.

Secondly, the Einsteinian science is connected to the metaphysics more by suggestion than argument. Whether we take the 'substance' or the 'self' approach, identification of Einsteinian spacetime with the one and only substance posited by monism appears to work on the basis of analogy, and thus not to be logically compelling. Natural scientists working within the Einsteinian paradigm certainly do not need to use the metaphysics of monism in their work. Even if they would agree that attributes need a bearer, it is the attributes of space-time which are their concern, and the issue of whether they have an accurate understanding of those attributes.

Some scientists may in fact be monists, in their view that the cosmos is the only substantial reality which we need to posit. But there certainly appear to be many competent scientists who believe that at least one other substantial reality needs to be posited, namely a creative deity, standing outside the cosmos. Some of the latter have even argued that such a hypothesis is needed on scientific grounds, as the only adequate explanation of the existence of the specific laws of nature which science has encountered.[18] Other scientists claim that a cosmos can be self-standing in this respect. Thus, scientists *qua* scientists are divided on the issue of how to conceptualise the cosmos.

There are, of course, arguments of a directly metaphysical nature, such as those of Spinoza, for saying that if there is a substance at all – and there must be at least one bearer of attributes – then necessarily there can only be one.[19] That is, substance-pluralism is incoherent. It is not clear that scientists who debate the issue mentioned in the last paragraph are employing such metaphysical considerations. At any rate, there is certainly no scientific consensus as yet on the metaphysical status of the universe, even if individual well-informed scientists hold strong views on the subject. Here as elsewhere it is vital to remember the distinction between the claims which individual scientists believe to be justified and what has actually been established by science.

Thirdly, as many have been quick to point out, ecologism's attachment to scientific explanations, while strengthening its immediate appeal for a Western audience who might otherwise view it as solely a 'New Age' fad, has the serious drawback which attaches to all systems of thought with a high reliance on science. This, of course, is the fact that scientific theories are always open to refutation and replacement with what are viewed as more accurate ones.[20] Perhaps we will in due course come to reject Einstein's theories in favour of some new comprehensive theory which is non-holistic, perhaps even in significant ways atomistic. Perhaps Darwin got it wrong, and biology has to seek another view which does not sustain theories of interconnection and system so prevalent in the biological sciences.

However, these problems, although real enough, do not invalidate Mathews' general approach. Plainly, ecologism is inevitably going to involve some radical departures in thought as well as behaviour, especially in the West from which many of the most serious threats to the environment emanate. Monism is not central to our intellectual traditions, but it is not unknown either, and thus such a metaphysic, if well explained and shown to make sense of our predicament, cannot simply be dismissed as too alien and esoteric to be convincing to large numbers of people in our culture.

The fallibility of science, too, is not a conclusive reason for ignoring its importance to the aims of ecologism. At present science often lends great support to ecologism's claims concerning how living systems work, how they relate to physical forces, how they can be harmed and what the result of those harms is likely to be. The possibility that at some future point these scientific claims, and the theories underlying them, may be falsified, does not make it indefensible to turn to them for support now, when they seem well founded. It does mean that ecologism has to be fallibilist too, and recognise that it may be wrong about those of its claims which rest primarily on scientific foundations.[21]

However, as was noted at the start of this chapter, ecologism is primarily a moral theory. The relations between scientific fact and moral value are not simple, and it is entirely possible for moral judgements to survive the demise of factual claims to which they have often been related. For example, even if it should turn out that life-forms on this planet, contrary to what Darwinian theory supposes, do not share a common origin, and thus forms of interconnection between human beings and other life-forms posited by this theory do not in fact obtain, the moral judgement that we should respect those other life-forms will arguably be unaffected, even if specific arguments for it may have to be jettisoned.

Let us now turn to consider some of the more specific claims which emerge in the course of Mathews' discussion, for these will enable us to clarify the content of a defensible form of ecologism. Specifically, there are four areas of her discussion we may profitably consider – the meaningfulness of life; fullness of being and biodiversity; intrinsic value; egalitarianism. In this chapter we will consider the first two of these, which are the most directly metaphysical of the four themes. The theory of the intrinsic value of selves and of their moral equality will be discussed at the relevant points in Part Two, as they concern the moral theory of ecologism, or at least its metaphysic of morals.

The Meaning of Life

Mathews' fusion of science and metaphysics promises to provide a vindication of the holism central to ecologism while simultaneously dealing with what is seen by the critics of the scientific world-view as its fundamental deficiency, namely its apparent lack of any convincing account of how the cosmos can have meaning. The search for meaningfulness is what is referred to when people speak of 'spiritual' matters. Science is said to satisfy the intellect but not the spirit. It explains how things work, but not their point or purpose. Mathews, however, deals with this by claiming that the cosmos is not just a substance, the sole bearer of attributes, but is also a self, just in the same sense that human beings and all other living things are selves. Thus, it has a point or purpose internal to itself, namely its own existence and flourishing. The holism of the theory then relates every particular self, each of which is a self-sustaining system, to the whole of which it is a part. The overall purpose or *telos* of the cosmic self thus becomes integral to the existence and purpose of individual selves, including human beings.[22]

Our conscious grasp of this point, our recognition and acceptance of our interconnectedness with the cosmic self and with every other specific self, enables us joyfully to find our meaning and purpose. Our intrinsic relation to the rest of the selves around us, and our consequent care for them, embodies our attempt to participate consciously in cosmic meaning by maintaining the conditions of flourishing of every other self upon which our actions impinge. The life enjoined by ecologism is a meaningful, and thus spiritually rewarding, one.

Within this framework we can make sense of destruction and death, which our knowledge of the Earth's biosphere everywhere reveals to us. Death, embodied in the play of natural selection, is the indispensable shaper of selves. Without killing, there would be no development or creation of more complex life-forms, such as ourselves.[23] We are in a position to realise that all this death and destruction around us has a positive tendency. Life and death do not cancel each other out, leaving a neutral overall situation of pure cyclicity, as some eastern religions may lead us to suppose. Rather, there is a clear direction which manifests itself in the increasing variety and complexity of the selves which are brought into existence. Creation overall prevails over destruction. If we do not grasp this point then mortality becomes a paradigm of the meaningless. It should instead be seen as the mechanism whereby the cosmic self unfolds itself by developing its potentiality for internal differentiation into actuality.

Following on from this, one might argue that this directionality is essential to the full differentiation of the cosmic self into all the specific selves which it is possible to have, provided that we assume that less complex forms cannot readily coexist with more complex ones in the same ecological niche, and that the more complex drives out the less complex. If so, then the only way in which the simpler and the more complex can both be manifested is in succession, with the more complex being formed from the less by the continuous operation of uniform causes, which obviates the necessity for separate creative events.

It is also clear that this claim enables us to take a positive view of the record of major extinctions to which life on this planet has been subject. These extinctions, the most recent one prior to the arrival of human beings being the asteroid impact which destroyed the dinosaurs and much else on Earth, have led to subsequent evolutionary explosions which have produced a more diverse biosphere and a more complex set of life-forms. The fact that these extinctions are random, and thus meaningless in the sense that there appears to be no intention directing them, does not detract from the fact that they have played a crucial role in the development of ever more complex selves. They thus acquire significance even though they do not emanate from a directing intelligence. (The difficulty of viewing the current human-caused extinctions in this light is raised in the next section.)

Has Mathews successfully shown how ecologism can present the holistic view of the universe which is so central to its vision without subordinating human life to a system which is intrinsically meaningless, thereby depriving human beings of the possibility of finding adequate meaning within their lives?

In spite of its great ingenuity and appeal, there are serious drawbacks from the point of view of ecologism to Mathews' metaphysical theory of monistic holism in this regard. Monistic holism can only provide a context of meaning if the cosmos is a self, and not just a substance. However, a self, on Mathews' understanding of the term, does not have to possess consciousness, let alone self-consciousness or discursive intelligence.[24] Arguably, however, even if the cosmos is a self, holistic connection with it will not confer meaning on the lives of human beings unless it is not only self-conscious but also conscious of, and caring with respect to, them and the conditions of their flourishing, in the manner of traditional deities. To be holistically connected with a non-conscious and uncaring self, even as an element of its self-development, is a situation which may intelligibly be regarded more as a nightmare than as meaning-conferring.

Of course, the cosmic self, as described by Mathews, is not completely devoid of points of similarity with human selves. Both are self-preserving and self-developing systems. However, human beings are not just selves, not even just self-conscious selves, they are persons with emotions, cultures and moral and aesthetic sense. The cosmic self appears to have none of the trappings of personhood. Nor is it clear what that would involve for such a being, or how we could recognise that it possessed such properties. This leads on to the fundamental point that personhood is essentially plural. Persons only become persons, develop as persons, in interaction with other persons.[25] If the cosmic self is a person, with which other persons is it interacting?

The attempt to deal with these problems may lead on to the identification of human persons with the cosmic self, so that the solution to the question of whether or not the cosmic self is a conscious person is to say that emergence of human personhood (and perhaps other persons in other parts of the cosmos) just is the cosmic self emerging into consciousness. However, whatever the attractions of this view might be, it is unclear how it can help with the problem of meaning of human life. For the problem is precisely that we are searching for meaning. The traditional theistic solution explains why we have this problem – the meaning is in the mind of God and must, therefore, remain inscrutable for us for a time until we pass to a different level of existence where all will be made clear. This neatly assures us of meaningfulness while deferring the solution of what the meaning is to a point beyond this life. But if we are searching for meaning it is scarcely reassuring or enlightening to be told that it is to be found within our consciousness because we are the cosmic self become conscious.

This explanation is also difficult to differentiate from the traditional humanist solution to the problem, which is to claim that there is no single meaning to life and human existence, but that we each find our own meaning within our lives (at least if we are lucky), albeit with certain common features deriving from our shared human nature. There is no 'beyond' or 'behind' where the 'true' meaning is to be found.

The role of the metaphysics of the cosmic self thus becomes very problematic on the hypothesis that that self's consciousness is identical with the consciousness of humanity and other persons. On this interpretation, Mathews' metaphysic appears to collapse into the scientific humanists' view, that the universe as a whole is meaningless and purposeless, and that meaning is something to be found only within human life, deriving from our particular natures. Meaning is thus not a bottom-level phenomenon, present in the universe, at least implicitly, from the beginning and permeating the whole of reality. Rather it is an emergent phenomenon, restricted in scope, and possessing no universal validity.

A further danger to ecologism is posed by the view that the cosmic self is developing progressively, via such phenomena as the wholesale destruction of life-forms in the major extinction episodes. If evolution is tending in a progressive direction, in the sense of moving, albeit in fits and starts, towards greater complexity, and if human beings represent the emergence of the cosmic self into conscious self-awareness, then it is hard to see what objection there can be to human beings pursuing the tendencies of their natures, creating culture out of calculated acts of destruction, and even bringing about another mass extinction of life on this planet.

Perhaps this is the way in which a new species of super-persons will eventually emerge, and the development of the cosmic self will be ratcheted up another notch or two. It is relevant to mention in passing one view which is a variant of this one, namely that of Easterbrook. He speculates that human beings have emerged precisely to remedy the defects of non-conscious nature – to put an end to disease, asteroid impacts and so forth and to remedy other defects, such as predation, by intelligent interference with the genes of other life-forms and to find alternatives to the wasteful, messy, violent and unpredictable mechanism of natural selection for our development into super-persons. Our role is to perfect nature.[26]

We have seen, then, that for ecologism there is much that is problematic about holistic monism, even interpreting the cosmos as a self. Ecologism certainly does need the key claims that human beings are interconnected with other life-forms and ecosystems on this planet, for this is central to the naturalist project of the philosophy. It needs also to argue for the claims that this interconnection is bound up with our having certain important moral obligations hitherto not properly recognised, that we need a new or perhaps renewed sense of humility with respect to the wider world, and that we must accept restrictions on the amount and kinds of things which we can do. However, the issue of how all this connects to the meaning of human life is very problematic. Draw an insufficiently close connection between the cosmos and human life, and meaning becomes impossible to achieve. Draw too close a connection, and humanity's present activities appear to receive validation, not rebuke.

In the light of these points, ecologism would do best to view meaning as an emergent property, not as something written into the very foundations of the universe. In other words, meaning should be understood as emerging from out of meaninglessness in the course of the interaction between conscious selves which are persons and the environment in which they operate, including other persons and other selves in Mathews' sense of the term. However, ecologism must heighten humans' awareness of their interconnection with that immediate environment, the Earth's biosphere, and to do so by tracing its importance not just for their material well-being, which appears still to be undeniable, but also for all those ingredients, especially the moral, the aesthetic and the emotional (especially the emotion of love in all its forms), which are essential to their finding meaning within their own lives.[27]

To repeat a central motif of this book, ecologism is primarily a moral philosophy. Although the nature of reality and scientific theory cannot be said to be irrelevant matters for moral theory, the connection between them exhibits a looseness of fit which means that political ecologists need to have a willingness to keep an open mind about which metaphysical and scientific theories lend most support to their views, or chime in best with them.

The standpoint of traditional religion, especially the various 'stewardship' positions which many claim are an important, though in recent centuries neglected, part of the outlook of the great world religions (although such positions are not the exclusive preserve of religions), is not necessarily to be rejected or attacked by those who hold this viewpoint. However, supporters of ecologism will always have a certain justifiable suspicion of the soundness of these religious perspectives as a basis for an environmentally enlightened world-view. The focus of most of these religions, especially Judeo-Christianity and Islam, is in the last analysis beyond this world. This is always going to provide scope for religious believers, as exemplified by John Muir's father, to develop the view that a concern for the natural world is itself a kind of blasphemy, given that our main concern should always and at all times be with the life which is to come.[28]

Fullness of Being and Biodiversity

Mathews provides a metaphysical reason for human beings to do all they can to preserve biodiversity. This is the claim that the self-realisation of the cosmic self requires it to realise within its self-development the maximum number of individual life-forms. It seeks, in other words, plenitude or fullness of being.[29] Human alignment with this cosmic aim thus requires our avoiding any activity which leads to the diminution of the number of different life-forms present within the Earth's biosphere. Should this aim of plenitude inform the philosophy of ecologism? Should ecologism criticise on this basis the undoubted fact that human beings have caused, and are presently causing to an increasing degree, the extinction of other life-forms on this planet?

Given the problems we have encountered in the previous section with the metaphysics of the cosmic self, it is clear that ecologism cannot readily support the idea of plenitude on that metaphysical basis. But more generally, there are great difficulties in using an ideal of plenitude to underpin a concern for biodiversity. The ideal of plenitude is a maximising ideal – we are to aim to promote the maximum number possible of diverse life-forms. In trying to understand this idea we need to be aware of some complexities. The first is that different possible life-forms may compete for the same ecological niche in such a way that only one can successfully occupy it at a time. Hence, if both forms are to be realised during the unfolding of spacetime, then they must do so at different times, with one form coming into existence, flourishing for a time, and then becoming extinct, leaving the niche to a new and different life-form.

The matter, of course, is made even more complicated by the fact that new niches may be brought into existence by the activities of specific life-forms, in the highly interactive manner characteristic of ecosystems. This may involve the destruction of existing niches, and thereby the destruction of the opportunity for possible, but not-yet-realised life-forms to occupy those niches.

All these phenomena have pervaded life on this planet for aeons before the arrival of human beings. It is impossible to show that natural processes have succeeded so far in achieving the maximum number of possible life-forms over the aeons before we humans got here. To show that that had occurred we would first have to work out the number of possible life-forms which could emerge from a certain clade[30] – for example, the number of possible different life-forms which could emerge from the genetic space occupied by frogs – holding the ecological niches occupied by the clade constant. Then we would have to work out the possible results for frog-diversification when the niches are altered in different ways, including what happens to the niches as the result of the emergence or extinction of other species. We would have to do this for all the clades of which we have (incomplete) knowledge. In effect we would have to trace the impossibly huge number of possible shapes which the tree of life could take and then show that the actual tree is the one which has (and will in future) produce the most branchings. One small area of one possible tree may show fewer branchings than a similar small area of another possible tree, but overall the former may contain more possible branches and thus better satisfy the ideal of plenitude.

Since this is an impossible task, we cannot be in the position to say that actual natural processes as they have operated in the absence of human beings are the best-adapted to embody the ideal of plenitude. We are, therefore, also not in a position to claim that what human beings are doing to the Earth's biosphere is detracting from the embodiment of that ideal. Hence, that ideal can have no part to play in the philosophy of ecologism. The case for avoiding anthropogenic extinctions must rest on a different basis than the ideal of plenitude.

There are, of course, arguments for avoiding the extinction of other species and the destruction of such holistic entities as ecosystems which are essentially prudential. The possible benefits we may obtain from these phenomena considered purely as resources are often cited as reasons for taking good care of them.[31] Their aesthetic, spiritual and recreational value for human beings now and in the future are encompassed in this prudential approach. It is also possible to make the powerful claim that human activity as manifested since the Industrial Revolution promises not to open up new possibilities for evolutionary diversification, but rather so to impoverish the biosphere that life-processes cease entirely on this planet. Our activity is constant and ever more invasive. In this respect we are unlike the previous causes of mass extinctions on the Earth which, however catastrophic they were, had a finite impact, allowing evolution to resume.[32]

These are strong arguments, although, based as they are on scientific theorizing, they are not without their detractors, some of whom claim that there is no evidence that human beings are actually having the deleterious effects to which supporters of ecologism, and many scientific ecologists, constantly refer.[33] However, ecologism's case for preserving the variety of species and ecosystems we actually have is a moral one, and derives from the moral considerability of each species and ecosystem taken separately. That is, it is not variety as such that is morally important from the point of view of ecologism, for taking that as a moral goal could in principle justify the human interference with natural selection to produce more new species (provided, of course, we did not thereby produce the elimination of more species and ecosystems than we created, thereby diminishing variety).

Thus, protecting from human-caused extinction the actual species and ecosystems we have is not a matter of forming an arbitrary preference for the present bundle of these things rather than some other, equally varied, future bundles, any more than seeking to preserve actual human beings is to exhibit an arbitrary preference for the present set over some future set which could come along in their stead. It is a matter of giving proper moral consideration to those things which are appropriate objects of it.

Variety as such is not devoid of certain kinds of value. Arguably the greater the variety of species and ecosystems within the Earth's biosphere, the more robust and stable the biosphere is, which is a potentially strong prudential point in its favour.[34] The sheer variety of life-forms and ecosystems adds to the variety of human life – the 'spice' of life, as the adage has it. More options are opened up for human beings to intertwine their life activities with the natural world if that world is as varied as possible. To the extent that these are important factors for the furtherment of human happiness they furnish an important moral consideration for preserving the widest possible extent of natural variety. This is, however, a matter of having a moral responsibility to human beings with respect to variety, rather than a matter of according variety in itself independent moral weight.

We may end this chapter by concluding that the main aims of ecologism with respect to the Earth's biosphere and its inhabitants rest on a moral rather than on a clear metaphysical basis. The contents of that moral theory, when the non-human world is granted moral status, do strongly rely on facts about interconnectedness – primarily physical, but also cultural and psychological – between human beings and other life-forms. Science, especially biological science, furnishes much of the necessary empirical basis for fleshing out human beings' specific moral obligations to the non-human world. But the moral case is one which could survive changes in our scientific theorising – if, for example, some of the forms of interconnectedness we presently suppose to exist were to be found not to hold after all.

We noted at the start that ecologism is strongly opposed to the atomistic metaphysics which has been so all-pervasive in our culture for so long. Holism and interconnectedness are the prescribed alternative metaphysical premises needed to reconnect us with the non-human world. What has emerged in the course of this chapter is that although interconnectedness as such may help us to recognise more easily than does atomism the moral claims of that world upon us, whether it actually does so or not depends very much on how we conceive of that interconnectedness.

It must be a way of seeing our connection with the non-human which at least does not detract from the moral status of the latter, and which preferably gives it support. This suggests that holism, in the last analysis, is not essential to ecologism. Acceptance of holism may have played a key part in forcing us to pay attention to the moral claims upon us of the non-human, but having got to that point, we can now see that even were atomism to be re-established in metaphysical and scientific theorising, the moral status of the non-human would not thereby be compromised.

Notes

1. F. Mathews, *The Ecological Self* (1991).
2. I have in mind here the doctrines concerning the identity of human selves with the wider 'whole' or 'self' of nature to be found in Arne Naess, *Ecology, Community and Lifestyle* (1989), Bill Devall and George Sessions, *Deep Ecology* (1985) and Warwick Fox, *Toward a Transpersonal Ecology* (1990).
3. Mathews 1991, chapter 1, especially pp. 31–40.
4. J. C. Smuts, *Holism and Evolution*, New York, 1926, pp. 86–7; quoted by Mathews 1991 on p. 94.

5. Mathews 1991, pp. 93–7.
6. See Mathews 1991, pp. 98–104 on self-maintenance and self-realisation.
7. Mathews 1991, pp. 25–9.
8. Mathews 1991, p. 108.
9. Mathews 1991, pp. 60–70.
10. Mathews 1991, pp. 77–90.
11. Mathews 1991, pp. 88–90.
12. Mathews 1991, pp. 57–8.
13. Mathews 1991, pp. 115–16.
14. James Lovelock, *Gaia: A New Look at life on Earth* (1979).
15. Mathews 1991, pp. 117–19.
16. Mathews 1991, pp. 147–63.
17. Mathews 1991, pp. 48–9.
18. See, for example, Paul Davies, *God and the New Physics* (1983).
19. As Mathews 1991 explains on p. 82, in part 1 of his *Ethics,* Spinoza tries to show that a substance necessarily exhibits the formal property of unity – there can only be one.
20. In chapter 4 of his *The Green Case* (1992), Stephen Yearley has an illuminating account of other problems posed for environmentalism by its reliance on scientific enquiry.
21. The need for ecologism to be fallibilistic has been urged by Michael Saward in his 'Green democracy?', in Andrew Dobson and Paul Lucardie (eds), *The Politics of Nature* (1993), p. 77. The implications of this for ecologism's views on democracy are discussed in Chapter 7.
22. Mathews 1991, p. 155.
23. Mathews 1991, pp. 153–4. Similar arguments have been put forward by various writers, such as Holmes Rolston in his *Environmental Ethics* (1988), pp. 239–43.
24. Mathews 1991, p. 104.
25. I discuss the nature and importance of personhood more fully in my 'Ecocentrism and persons' (1996).
26. Greg Easterbrook, *A Moment on Earth* (1996), chapter 10. As John Barry has pointed out to me, this idea has ancient antecedents, as explained in John Passmore, *Man's Responsibility for Nature* (1980), chapter 2.
27. The conclusion reached here bears some resemblance to that of Rolston 1988, chapter 9. Rolston there puts forward the idea that meaningfulness may be derived from one's participation in a narrative. Thus, the story we are now in a position to tell of the natural history of the cosmos and of our own emergence within a context of natural forces and processes which produce value for the entities so created, provides just such a narrative to embed a sense of the meaningfulness of our own being. This may be so, he suggests, even if our own emergence has, according to the story, no inherent necessity

(p. 345). Clearly, intelligibility and meaningfulness are closely connected.

28. This view held by Muir's father is recounted in Frederick Turner, *John Muir: From Scotland to the Sierra* (1997), p. 69.
29. Mathews 1991, pp. 127–8.
30. A clade is a lineage of species the members of which evolve out of those preceding them.
31. Wilson 1992, chapter 13, is eloquent on these factors.
32. See Aitken 1996 for the development of this point.
33. Greg Easterbrook shows particular scepticism respecting these claims. See his *A Moment on Earth* (1996), chapter 30.
34. However, the relation between diversity and stability is still a matter of dispute within environmental science – see Michael Allaby, *Basics of Environmental Science* (1996), pp. 169–73.

3

Biology and Ecologism: The Case of Sociobiology

A CENTRAL TENET OF ECOLOGISM is that human beings are to be understood, as are other animal species, primarily as a part of the natural world. The nascent scientific discipline known as sociobiology also insists on seeing *Homo sapiens* as an integral part of the natural world, to be grasped in theoretical terms as one species of social animal, of which there are many other examples – the social insects, the social mammals such as wolves and the social primates. Famously, or notoriously, according to taste, this position led the leading proponent of sociobiology, Edward Wilson, to claim that the social sciences – sociology, anthropology, psychology, geography, political science, economics, history – ought in principle to be reconceived as branches of that part of biology which seeks to elucidate the biological basis of human social behaviour.[1] This view of Wilson's will receive some support from considerations put forward in Part Three of this book.

A key claim of sociobiology is that the broad outlines of human social behaviour, covering human morality, religion, politics, economics and social structure, can be understood and explained in terms of neo-Darwinism. That is, they should be seen as having been produced under the influence of 'epigenetic rules', the genes for which have emerged as the result of the normal workings of natural selection upon the human species.[2] Thus it ought in principle to be possible to determine at least the limits of possible social behaviour for the species *Homo sapiens*, even granting, as Wilson does, that much of actual human social behaviour is mediated by cultural structures which do not have any direct genetic foundation.

An important implication of this picture of the epigenetic basis of human sociality is that human beings may be able to produce, by using their intellectual powers, prescriptions for forms of human social life which they are unable to put into practice because those forms of social life are not ones within which members of our species, given the epigenetic rules which govern them, are able to live.[3] Wilson mentions, but does not explore fully, several examples of such conceivable, but unliveable, social arrangements, such as Plato's ideal republic, some of the child-rearing practices of kibbutzim in the early days, pure *laissez-faire* societies and anarchist communes.[4]

A question which arises for ecologism out of these claims is whether the preferred forms of social organisation mooted by ecologism fall into the category of intelligible, but unliveable, forms of social life for our species. Two issues are involved here. First, and fundamentally, there is the issue of whether human beings are capable of accepting ecologism's crucial claim of the moral considerability of the non-human. If this can be shown by sociobiological theory to be an impossible ideal for members of our species, then ecologism will be faced with the irony that the theoretical perspective which prima facie best does justice to its naturalist position explicitly rules out the form of human social life which its moral and ecological arguments prescribe.[5]

The second, less fundamental, issue is whether sociobiology does rule out as unviable for human beings any of the specific forms of social organisation for which political ecologists wish to argue. Wilson's rejection of pure *laissez-faire* will be music to the ears of most political ecologists. But his case against anarchism will not be welcome to many political ecologists who see the dissolution of the state and its replacement with some form of anarchist-commune organisation as essential to the achievement of ecologism's aims.[6] In this chapter I will concentrate on the former, fundamental, issue. The question of the preferable mode of social–economic–political organisation for the attainment of ecologism's aims is one to which we will return in Parts Three and Four.

Among environmental ethicists, Callicott is notable for claiming that sociobiology 'offers a tremendous resource for moral philosophy in general and for environmental ethics especially'.[7] For others, sociobiology is an intellectually unsustainable and morally objectionable project. It has come to be viewed by many as inherently committed to reactionary prescriptions for social life, based on a picture of human beings as competitive and almost entirely self-interested. It has been criticised by Benton, among others, as seriously flawed, both as biology and as an approach to the understanding of human culture and society.[8]

Biologically speaking, its attempts at reductivist explanations of everything about organisms in terms of the neo-Darwinian theory of natural selection as applied to genomes are said to fail even in principle to explain a wide variety of biological phenomena, such as the developmental differentiation of cell structure. Such phenomena are said to require holistic or 'field' explanations, which introduce explanatory levels which cannot be reduced to the direct effect of the genomes alone. There are other complicating factors, such as 'genetic drift', which are said to show that even the genetic structure of organisms cannot be fully accounted for by the neo-Darwinian explanatory paradigm.[9]

As applied to human beings, the theory faces further serious objections of principle, such as the following. The environment with which the human organism interacts has for long been one modified precisely by human social activity, hence the latter must figure as an explanatory factor in any attempt to account for human biological evolution. Such a social environment has developed increasing independence from the purely biological sphere. Thus the habitat which is supposed to exert selective pressures on the human genome is itself to an increasing degree an autonomous product of human social activity. Human culture, therefore, does not simply mediate between human biology and human behaviour, it forms an autonomous level which cannot be reduced to biology and which plays an irreplaceable explanatory role in accounting for human behaviour.[10]

It may be, as many of sociobiology's critics aver, that human social life is so overwhelmingly a cultural rather than a genetic phenomenon, that no branch of biology can, even in principle, throw any light upon the factual and normative issues with which human beings wrestle in the process of reaching self-understanding. However, there is the clear danger for ecologism in pursuing the critique of sociobiology as far as possible that it ends up by supporting the dualism between humanity and the rest of the natural world, the denial of which, as we noted at the start of this chapter, is supposed to be one of its central, defining, claims.

After all, some distinctive implications for human society are supposed to follow from ecologism's naturalistic perspective. If we make human society too autonomous of the natural environment nothing, or at least nothing very important, seems to follow from the undeniable fact that we are an animal species among others. On the other hand, the advantage of seeking to establish a looseness of fit between human beings' biological nature and their culture is that it gives ecologism a conceptual space within which to explore a variety of forms of social–economic–political structures in pursuit of its normative goals. By contrast, if culture reduces to biology, biology reduces to neo-Darwinism, and neo-Darwinism depicts us as an aggressive, self-centred species, then the moral and social theorising of ecologism are in grave danger of becoming mere hot air. Neo-Darwinism may not, of course, endorse such a bleak picture of humanity, but it may appear to be a risky strategy to put all one's eggs in the neo-Darwinian basket.

We now possess, however, many studies of human behaviour from a biological perspective. We also have available many studies of non-humans, especially of our nearest relatives, the great apes, which examine their social lives for the presence of ethical, or proto-ethical, forms of behaviour. These give strong support to the idea that there is an identifiable genetic basis to our ethical and social norms.[11] This is evidenced in the fact that, as Singer has emphasised, some kinds of moral rule are fundamental to every human society, in spite of manifest cultural variations between them. Further, similar kinds of norm appear to be detectable in the social lives of other species, especially chimpanzees. These norms include:

(1) obligations on members of a family to support their kin;
(2) obligations of reciprocity, in return for favours done and gifts received;
(3) constraints on sexual relations.[12]

This chapter will prescind from the issue of whether sociobiology is good biology, and whether we can establish a *rapprochement* between biological and social scientific understanding sufficient to abolish pernicious dualisms, but not so close as to rule out free play in the design of human social arrangements. Instead, it will address the question of whether sociobiology, as it has been developing since the 1970s, does rule out ecologism's moral, social and political prescriptions. This will reveal whether ecologism has a stake in the failure of sociobiology or whether it may still hope for the success of some, suitably refined, form of sociobiological theory so as to provide a secure naturalist foundation for human social thought in general and its own prescriptions in particular.

Let us, therefore, consider the question of whether sociobiology's epigenetic rules do rule out the political prescriptions of ecologism. We may usefully begin with an examination of the ideas of one theorist who has attempted a development of sociobiological ideas with direct application to human beings, namely Michael Ruse in his *Taking Darwin Seriously* and *Evolutionary Naturalism.*[13] Ruse has offered in these works a sociobiological interpretation of the phenomenon of human morality. This interpretation prima facie threatens the viability of ecologism's central ethical position, which rejects chauvinist anthropocentrism, and has direct implications for political morality.

I will bring his speculative arguments into contact with those of Edward Wilson, whose concept of 'biophilia' as a genetic predisposition of members of *Homo sapiens* appears to give some hope that an ethics which attributes moral value to the non-human is not a complete non-starter for human beings.[14] I will try to develop Wilson's claim that there is good reason to suppose that natural selection could have selected for the biophiliac impulse in species such as our own. I will also try to show that, even within the theory of epigenetic rules developed by Ruse, there is reason to suppose that human beings can espouse a non-anthropocentric ethics which possesses a strong motivational character.

Ruse's basic claim concerning morality as a human phenomenon is that it is a genetically based tendency to behave in a way which as a matter of fact enables those human beings who possess it to reproduce themselves more successfully than those who do not.[15] There is no point or purpose to morality other than this. Its specific function among human beings is to secure co-operative behaviour among them, thereby improving the chances of individuals to reproduce. At its heart is the phenomenon of altruism, which it has always been a challenge for Darwinism to explain, involving as it does the putting of the interests of others ahead of one's own interests. The altruistic elements in morality are based primarily on the phenomena of kin selection (which has been used extensively to explain, in terms of neo-Darwinism, the self-sacrificial activities of the social insects) and reciprocal altruism – mutual back-scratching – which the shrewd, socially attuned brain of humans has, in part, evolved to engage in.[16]

However, although these two phenomena are the fundamental

ingredients in human morality, Ruse argues that the characteristic sense of compulsion and objectivity which human beings typically experience with respect to moral rules is also necessary. It is needed if morality is to do its evolutionarily prescribed task of overcoming individual self-interest so as to secure the benefits of co-operation, thereby enhancing reproductive success. The sense of compulsion and objectivity is, however, an illusion.[17]

This implies, on the face of it, that human morality can only easily be directed towards other human beings, or at least to other beings able to engage in co-operative relations, mediated by moral concerns, with human beings. Non-human organisms do not meet this requirement. It also implies something which Ruse is at pains to emphasise, namely that even between human beings, morally motivated behaviour will be very limited in its motivational power. It will be most strongly manifested towards those individuals with respect to whom kinship relations come into play and to a lesser degree with respect to non-kin with whom reciprocal altruism may be expected to pay co-operative dividends. The sense of moral obligation towards human individuals who are beyond the reach of either of these forces will be very weak – at most the sense of obligation will be held as a purely intellectual position, playing at best only a faltering and intermittent part in human beings' moral lives. This, Ruse contends, is borne out by what we observe of human moral conduct and by what we can readily sense, if we are frank with ourselves, via introspection.[18]

This theory has various advantages, Ruse contends. It explains why all moral justification must come to an end at some point, and shows that that point is precisely the one at which we can cite the putative epigenetic rules as explanatory concepts. It enables us to avoid the pitfalls of relativism – these are species-specific and species-wide epigenetic rules – and demonstrates why the attempt to adopt a position of moral scepticism, perhaps on the basis of our grasp of precisely this sociobiological theory, is only at best an intellectual position, not an existential one. For we cannot escape the effects of the epigenetic rules even when we understand the mechanism underlying those effects and see that morality has no objective basis.[19]

From the point of view of ecologism, however, it has the great drawback that it appears to rule out the possibility of any non-anthropocentric moral theory's having motivational impact to any degree on human beings. It therefore makes the political prescriptions of ecologism, which rest upon the granting of moral considerability to the non-human, appear to be wholly non-viable from the start, whatever other, more practical, difficulties they may also face.

However, there may be found within Ruse's own ruminations on the epigenetic rules which apply to human beings some resources with which to construct an alternative position, one which suggests that a sense of moral obligations to both detached human beings and to the non-human may have a reasonably powerful motivational force.

As Ruse himself is at pains to point out, human morality can 'soar above' biology under the influence of human culture.[20] This suggests that kin selection and reciprocal altruism in the human case are only the basis for human altruism, not its entire content. It seems possible that once the idea of 'respecting the interests of others for their own sake, whether it pays or not' is embedded in the human mind, the human intellect can use it as a starting point for developing the fuller implications of the idea of altruism. However, Ruse's arguments would lead us to suppose that such a purely intellectual construct would have limited motivational impact on us. Altruism towards strangers and concern for the Earth's environment will be bound to have less 'pull' on us than concern for kin (and, perhaps, compatriots) or for our local environment. Ruse's argument is in part intended to highlight and explain our limited altruism in these areas, thereby giving us a clear grasp of why it will require a special effort to extend human moral concern into them.[21] However, the fact that he relies on a theory of epigenetic rules implies that the prospects for such an extension look rather bleak.

We may reasonably acquire a more optimistic view on this matter if we turn to another part of Ruse's theory. This covers the topic of Darwinian epistemology, with respect to which he argues that we are governed by epistemological epigenetic rules which are embodied in the rules of logic and such common-sense requirements of reason as consistency.[22] He does not say so himself, but by parity of reasoning with the example of human ethics, it appears to be a requirement of the idea that such rules improve the reproductive success of their possessors that they be accompanied by a powerful motivational element.

After all, the principles of scientific thought, which are commonsense writ large, are no use unless we are motivated to follow them – a 'will to knowledge' perhaps. Thinking in a controlled and concerted way in accordance with the rules of logic is often as difficult as, if not more difficult than, thinking in terms of the requirements of morality, particularly when, unlike the usual moral case, there may be no immediate practical import to the thinking. We appear to need, then, to postulate on an epigenetic basis a set of characteristic human epistemological feelings – curiosity; desire for knowledge; enjoyment of enquiry, argument and debate; dislike of inconsistency, in oneself or others; love of the 'universal standpoint of reason'. Individuals with such feelings and the epigenetic epistemological rules will be likely to do better than those who do not.

The suggestion may now be made that the urge towards the universal standpoint embodied in objective knowledge may be of sufficient motivational force in most human beings to counter the motivational force of the purely moral epigenetic rules which, if Ruse's analysis is correct, tend more towards the local rather than to the universal. Avoidance of inconsistency is clearly a requirement of both theoretical and practical reason. Thus, part of the 'training' in moral thinking which Ruse acknowledges to be essential in the human case will precisely be to employ the canons of rational thought in the moral context. This will produce simultaneous moral and intellectual advance, and harness the strong motivational force which may be presumed to attach to the epistemological epigenetic rules to the purposes of moral thought.

This suggests, then, a cautious optimism about the possibility of reconciling a Darwinian approach to the explanation of moral thought with the environmental ethic of ecologism.

Finally, let us consider a further possible source of motivational force, explicable in Darwinian terms, to underpin ecologism's environmental ethic, namely what Edward Wilson has called 'biophilia'. In his discussion of human nature from the sociobiological viewpoint, Wilson suggests that there are various values to which we are predisposed by our genes.[23] He designates as the 'primary' ones a sense of the nobility of our species, a commitment to the preservation of the whole human gene pool and universal human rights, based on 'mammalian' individualism (as contrasted with the self-effacement of the social insects). It is notable that in this list we are presented with values whose orientation is clearly moral and which appear to be universal in scope. There is no sign here of the parochialism which Ruse suggests is inherent in the kinship/reciprocal altruism basis of moral thinking among human beings. Wilson, who after all is one of the sources of the kinship/reciprocal altruism theory, appears at this point at any rate to be assuming something like the account we have given above of the way in which moral thinking has a strong tendency in human beings to move towards the universal.

'Secondary' values encompass enthusiasm for exploration, exaltation from discovery (examples of epistemological emotions?), triumph in battle/competition, satisfaction from 'altruism', pride in ethnos/race, strength from family ties and biophilia for other life-forms. Biophilia, therefore, is only a secondary value, and one among a number in that category.

Let us assume that Wilson has good reasons for these claims. What does this imply about the prospects for the acceptability of ecologism's preferred form of social life, in which the recognition of the moral status of the non-human is to be given a central place? Out of the ten values cited, only biophilia seems to form the basis for human motivation towards taking into moral account the interests of the non-human. The others appear either neutral with respect to this goal or else to tend to support a purely anthropocentric view of the world.

A further puzzle is that the examples which Wilson often cites as examples of biophilia are a rather curious hotch-potch which make the label he chooses for them seem inapt.[24] For example, irrational fear of snakes and spiders is said to exemplify biophilia, when it rather suggests biophobia. On Wilson's account there is an explanation of these fears which is to be given in terms of natural selection, and they are directed towards natural phenomena, but it is simply odd that they should be cited as examples of biophilia. Perhaps a term such as 'bioconnection' would be more appropriate, as indicating the key idea that we have co-evolved with other species in such a way that the latter have left deep psychological, and thus cultural, imprints upon us.[25]

Other examples are more intelligible in terms of the label, but their explanation in terms of natural selection is obscure. Such, for example, is his claim that human beings' preferred natural environment in which to live is a savannah landscape with dwellings placed on a high prominence in sight of water.[26] One can see that this is a kind of 'love' and that it is directed towards the kind of environment in which we are supposed to have evolved, but it is not clear why an animal whose main claim to fame as far as life-environment goes is that it lives almost everywhere should have internalised a preference for this specific kind of landscape. One would think that it sits rather ill with the putative value 'enthusiasm for exploration' cited above as one of Wilson's epigenetic values.

What ecologism would like to get from the sociobiological explanation of biophilia is a reason for saying that it improves the reproductive fitness of human beings if they acquire by natural selection the tendency to cherish and show moral concern for at least some important aspects of the non-human natural world.

Wilson does make claims which go some way in this respect.[27] He tells us that:

> From the scant evidence concerning its nature, biophilia is not a single instinct, but a complex of learning rules that can be teased apart and analyzed individually. The feelings molded by the learning rules fall along several emotional spectra, from attraction to aversion, awe to indifference and peacefulness to fear-driven anxiety.[28]

Such learning rules, even if weak in modern humans, have persisted, he claims, and have pervaded all of our culture. But the experience of modern humans represents a tiny fraction of that which has been dominant during human existence:

> For more than 99 per cent of human history people have lived in hunter-gatherer bands intimately involved with other organisms. During this period of deep history, and still farther back, into paleohominid times, they depended upon an exact learned knowledge of crucial aspects of natural history.[29]

However, what these considerations point to, even if well founded, is the idea that it has been important to human beings to know how to exploit their environment for their own benefit, how to 'find their way around it', what to seek and what to flee from. There is as yet no clear reason for speaking of a caring, loving or cherishing attitude as an indispensable element of biophilia. He does have some further ruminations which point towards the idea that our connection with other species has 'immense aesthetic and spiritual value for us'.[30] But what he then does is to urge upon us the importance of a richly biodiverse world for religious, cultural and more generally psychological reasons. Ecologism will be inclined to accept his exhortations. But, logically speaking, they are the wrong kind of claim to be making here. What is needed is some sociobiological argument to suggest that in some way these non-instrumental attitudes of interest, care and concern for the non-human world have an epigenetic character, so that ecologism's ethic can be shown to have a firm foundation, and thus motivational basis, in human nature.

The following speculation is in line with the general tenor of Wilson's argument, while seeking explicitly to find a genetic basis for the ethic of ecologism.[31] Given that we have evolved as explorers of environments and drastic modifiers of them, it is clearly a danger to our reproductive success that we treat the natural environment in a way which harms it, and thus probably harms us, in order to achieve short-term benefits. Human groups which came to the view that the non-human is worthy of respect and care, perhaps by deifying it, might take better care of it and thereby secure the wherewithal for life and reproduction better than groups which were not similarly inhibited. Of course, this hypothesis depends on the supposition that it does not, on average, pay to be short-term despoilers. But that seems a reasonable hypothesis, especially if despoilers are prevented from moving on to fresh pastures by geographical barriers or other human groups already in occupation of adjacent territory.

Thus, human groups whose individual members possess something like an epigenetic reverence for non-human nature, at least to a degree, may reproduce better than groups whose members possess a purely instrumental attitude towards it. The problem with the latter is that if it is the sole attitude prevalent in a human group then there is no obvious stopping-point built into it, especially in the absence of an extensive understanding of how the natural world functions. Evidence compatible with this hypothesis is provided by the oft-cited point, emphasised by Sylvan and Bennett,[32] that human beings appear always to have drawn the moral boundaries of their world in a way which includes some aspects of the non-human as objects of awe and reverence – to be cared for, placated and worshipped.[33]

Evidence also compatible with it may be drawn from the fact that even in our modern artefactual world, when individuals may live their entire existence in an almost totally artificial environment, people show a yen for, and exhibit moral attitudes towards, the non-human natural world. They make gardens, keep house plants and domestic pets, go for country walks, watch natural-history films and so on.[34] This provides some reason for supposing an epigenetic basis for this behaviour.

It might be urged that the attitudes just cited are purely cultural constructs, passed on from group to group by the normal means of cultural transmission, and surviving only because they are functionally useful for groups which adopt them. There would then be no reason to cite biophilia as their epigenetic basis. However, in a largely artefactual world such as prevails in the 'advanced' societies of the planet, it is not obvious what advantage accrues to human groups which socialise their members into the above-mentioned kinds of behaviour, except perhaps that it is in the economic interests of some groups – the suppliers of garden paraphernalia, pets, country tours and so on – that people should have such preferences. There are individuals who exhibit none of the above behaviour, who are indifferent to the non-human natural world, but these may be explained on either hypothesis, as genetically different or as imperfectly socialised. However, such individuals seem to provide as rich a market for products as the more biophiliac individuals just cited. It is not obvious that a society composed only of those indifferent to the natural world would suffer economically from that fact.

I conclude, therefore, that a plausible hypothesis can be put forward which suggests that there is a genetic basis for human attitudes of caring for the environment, in at least some respects, for its own sake. If so, then we may be more optimistic than we might otherwise be that there is an ingrained motivational basis to support some form of human society of the kind which ecologism is proposing, in which the non-human world is accorded moral consideration in human deliberations.

Ecologism and sociobiology may yet be allies rather than foes, therefore. If sociobiology does support ecologism's prescriptions, and can vindicate its Darwinian approach to the analysis of human nature, then this will have the great advantage for ecologism of helping to produce a thorough and coherent, descriptive and prescriptive, development of the fundamental thought of ecologism – that human beings are part of the natural world.

Notes

1. E. O. Wilson, *Sociobiology* (1980), p. 271.
2. Wilson explains that an epigenetic rule

> is any regularity during epigenesis that channels the development of an anatomical, physiological, cognitive or behavioural trait in a particular direction. Epigenetic rules are ultimately genetic in basis, in the sense that their particular nature depends on the DNA development blueprint.

(From p. 370 of C. Lumsden and E. Wilson, *Genes, Mind and Culture,*

Cambridge: Harvard University Press, 1981, cited in Ruse 1986, p. 143.)

3. As he explains in *On Human Nature* (1978), p. 18:

> Evolution has not made culture all-powerful. It is a misconception ... that social behaviour can be shaped into virtually any form ... Although the hundreds of the world's cultures seem enormously variable to those of us who stand in their midst, all resources of human social behaviour together form only a tiny fraction of the realised organisation of social species on this planet and a still smaller fraction of those that can readily be imagined with the aid of sociobiological theory.

4. For example, he says in *On Human Nature* (1978), p. 208:

> As our knowledge of human nature grows, and we start to elect a system of values on a more objective basis, and our minds at last align with our hearts, the set of trajectories [along which societies can evolve] will narrow still more. We already know ... that the worlds ... of the extreme Social Darwinist and the Anarchist are biologically impossible.

5. Of course, this is not the only naturalist position available to ecologism. For example, in Peter Dickens, *Society and Nature* (1992), a naturalist position based on Bhaskar's critical realism is outlined. Sociobiology is here considered because it makes the most explicit connection between the theoretical understanding of the human species and that of other species, under the rubric of neo-Darwinism. It thus helps to focus most sharply the issues concerning naturalism for a science-oriented ideology such as ecologism.
6. An example of such a view can be found in Alan Carter, 'Towards a green political theory' (1993).
7. J. B. Callicott, *In Defense of the Land Ethic* (1989), p. 11.
8. See Ted Benton, 'Biology and Social Science' (1991).
9. See Benton 1991 and Dickens 1992, chapter 4.
10. See Benton 1991, pp. 19–23.
11. A useful recent source of some of this writing, germane to the ethical and political issues central to ecologism, is the section 'Common themes in primate ethics', in Peter Singer (ed.), *Ethics* (1994). Excerpts from the work of biologically oriented students of human behaviour, such as Barash, Trivers, Symons and Axelrod, and of primatologists, such as Goodall and De Waal, reveal the promise of the naturalistic approach to the study of our own species.
12. Singer 1994, p. 56.
13. Ruse 1986 and 1993.
14. Wilson 1984.
15. The following summary of Ruse's argument is drawn from Ruse 1986, chapter 6.
16. Another suggestive presentation of a sociobiological account of the evolutionary origin of morality among human beings is found in

Callicott 1989, especially chapter 4. He there usefully emphasises the possible presence of proto-ethics in other social species, thereby helping to challenge the belief that morality is unique to human beings on this planet.

17. Ruse 1986, p. 221. It is possible to reject this conclusion of Ruse, however, by interpreting the very facts which he claims undermine the objectivity of human values as doing the precise opposite. H. Rolston's *Environmental Ethics* (1988), especially chapter 6, is a magisterial exposition of such a position. It offers a naturalistic account of value which cites the evolutionary story to show why human creatures should come to recognise the values which they do. His point is that nature creates by means of natural selection the values which living creatures recognise, consciously or otherwise, in their lives. Hence, their valuings conform to something existing outside of themselves. Of course, if 'objective' is supposed to refer to something existing beyond nature, then this will not work. But as a critique of the naturalistic ethics of Ruse it is effective in showing that such an ethical position can still have an intelligible place for the concept of objective value.
18. Ruse 1986, p. 240.
19. Ruse 1986, p. 252.
20. Ruse 1986, p. 223.
21. Callicott's solution to the problem of how human beings might be expected to extend moral concern to the non-human relies on the view, derived from Hume and Darwin, that we have an evolutionarily produced set of 'inborn moral sentiments which have society as such as their natural object' (Callicott 1989, p. 85). Once we have a grasp of ecology, which demonstrates that the human and the non-human do form a 'society', the human moral sense may be expected to take the biosphere as an object of moral concern: 'Therefore the key to the emergence of a land ethic is, simply, universal ecological literacy' (p. 82). Ruse's account has the merit of not allowing us to take this for granted.
22. Ruse 1986, chapter 5.
23. Wilson 1978, pp. 198–9.
24. See, for example, his *The Diversity of Life* (1992), p. 333.
25. He comes close to this view himself on p. 334 of *The Diversity of Life* where he characterises biophilia as 'the connections that human beings subconsciously seek with the rest of life'.
26. Wilson 1992, p. 334.
27. The account which follows refers to a recent essay in which he directly considers the moral importance of biophilia – 'Biophilia and the environmental ethic' in Wilson 1997.
28. Wilson 1997, p. 165.

29. Wilson 1997, p. 166.
30. Wilson 1997, p. 176.
31. This whole field is rife with speculation rather than established facts or theories. This is inevitable given the status of neo-Darwinism as a high-level explanatory hypothesis whose *explanandum* and *explanans* both resist easy experimental testing. However, from the point of view of this book, the key issues concern precisely what speculations Darwinism permits and what it rules out *ab initio*.
32. Richard Sylvan and David Bennett, *The Greening of Ethics* (1994), pp. 29–30.
33. This is also a theme emphasised by various contributors to W. Sachs (ed.), *Global Ecology* (1993) – discussed in Chapter 11 – who see the traditional practices of communes in India and elsewhere as embedding a concern for the maintenance of a healthy local environment within religious practices, such as the creation and maintenance of sacred groves, the leaving of land fallow during certain periods for religious reasons, and so forth.
34. Wilson makes similar points in support of the idea of biophilia as 'bioconnectedness' on p. 334 of *The Diversity of Life*.

Part Two

MORALITY

4

The Moral Theory of Ecologism

THE MORAL IDEAS WHICH have been developed since the 1970s by those who have contributed to the development of ecologism are among the most distinctive aspects of the ideology. The fundamental thought which guides ecologism's reorientation of moral theory is that (1) the non-human is worthy of moral consideration and that (2) this is in part due to its non-instrumental value – that is, the value of the non-human does not rest on its contribution to human welfare.

One variety of this 'fundamental thought' faces difficulties. This is the 'strong ecocentric' view, which grants moral centrality to the biosphere, and grants particular creatures, including human beings, only a secondary value. Prima facie such ecocentrism would seem to countenance the possibility that human beings could be morally required to commit mass suicide in order to protect the integrity of the biosphere, for we might turn out to be the equivalent of a planetary disease, like cancer.

Ecocentrists who wish to avoid this conclusion may seek to do so by highlighting the claim of interconnectedness between the human and the non-human. What benefits the biosphere benefits human beings. This claim is buttressed by many arguments which are variations of the basic thought that a fully adequate theory of what human beings are will be one which develops a concept of the human self as connected physically, biologically, culturally and spiritually with the biosphere of which it is a part.

On some versions of this view, such as the 'wider self' view of theorists such as Naess, Sessions, Devall and Fox,[1] the biosphere is fully integrated into the individual's sense of self by the identification of self and world. Arguably, moves such as these are needed not simply to avoid the unpalatable implication mentioned above, but also to motivate many human beings actually to attend to the well-being of the biosphere. For only if such human beings can be brought to believe that their well-being is at stake in the protection of the rest of the biosphere will they see any reason for them to exert themselves to protect it.

If we accept this kind of approach then it appears that in putting the biosphere at the centre of our moral concern we turn out always to be putting ourselves there too. The implications of this point have been recently clarified by Tim Hayward. He suggests that we use the terms 'human chauvinism' and 'speciesism' to refer to the view that only human beings count morally, and employ 'anthropocentrism' to refer to the view that human beings are of central moral concern.

The latter view, he argues, is defensible – indeed unavoidable for human beings – the former is not. The latter view is compatible with the claim that the biosphere, other species and other creatures are morally considerable also. They thus have to be taken account of in human moral deliberation and it is possible that their interests and well-being will on occasion trump that of human beings. But this is not to put human concerns and interests on a complete par with those of non-humans. The idea of moral considerability admits of degree and can readily extend to the non-human. We should now, therefore, claim that it is this view which is distinctive of ecologism's approach to morality, rather than the reduction of the moral status of human beings to that of any other species and/or to a position subordinate to that of the biosphere.

This view allows development of the case for the interconnection between human well-being and that of the non-human which, as we have noted, is an important part of the moral position of ecologism. It also avoids an insurmountable difficulty which emerges when attempts are made by proponents of 'strong' ecocentrism to elucidate what in practice is to be done in order to place the biosphere at the centre of moral concerns. This difficulty is, as Hayward puts it, that of the radical indeterminacy of the idea of the 'good' of the biosphere.[2] We have no way of telling what particular conformation of the biosphere is better for it. We can more readily tell which conformation is better for a given species, and most readily of all what is better for our own species. These points suggest, he concludes, that the correct value to assign to the biosphere is an instrumental one, which has the effect, of course, of removing the biosphere from the centre of moral concern (although this does not remove it from the ambit of morality altogether).

This is essentially the version of ecologism's moral position which will be defended in the remainder of this book. However, Hayward is sceptical of the idea of intrinsic value as applied to the non-human, and prefers to emphasise the thesis of interconnectedness between human and non-human well-being as a way of extending moral concern to non-humans. In taking this view he joins with a theorist such as Bryan Norton who has strongly pressed the issue of whether the aims of normative environmental theory in general, which clearly encompasses ecologism too (although this is not a term which he uses), can be attained without the need to resort to any distinctively new moral position. Let us, therefore, examine Norton's position.[3]

Does Ecologism Really Need a New Moral Theory?

Norton's main idea is that the crucial concept of interconnectedness provides a sufficient theoretical basis to afford other species, ecosystems and the biosphere all the protection they need. Recognition of such interconnectedness is precisely what the science of ecology engenders. What we should derive from ecological understanding is a diagnosis of a certain kind of human arrogance, namely the belief that human beings are able to break free of such interconnections, viewed as trammels on human development, and make the world over entirely in the light of human needs and desires alone.[4]

Once we have grasped the folly of such arrogance, and attained a just appreciation of the non-human context within which alone human life is possible, then we can quickly come to see the vital importance of thinking in terms of that context and of preserving the viability of natural ecosystems and the biosphere as a whole, for the benefit of ourselves and our posterity, for whom we are required to have moral concern. Ecological management will enable us, once we start thinking along the right lines, to implement the multifarious ways in which human activities can be integrated into the natural environment so as to protect both that environment and the sources of human material, cultural and spiritual well-being.

In practice, Norton argues, the development of a distinctive thesis concerning the moral considerability of the non-human does not lead its supporters to offer any policy recommendations different from those put forward by environmentalists who reject the thesis. He clearly, however, believes that the thesis can be an impediment to the acceptance of such recommendations, since the moral theory involved will strike many people as either laughable or repugnant or both. The emphasis on such a theory threatens, therefore, to forfeit the support of such people for policies which they could well be led to support if they were simply encouraged to think 'contextually' (to use his preferred term).

The approach which focuses on the interconnectedness of human and non-human revealed by science, rather than the putative moral status of the non-human, also has the great advantage of allowing people with a wide variety of value-commitments, some ostensibly at variance with those of the Green movement, to agree on policy prescriptions on the basis of their acceptance of the contextual approach. A simple example which he gives of this is an agreement between hunters and birdwatchers to fight to protect wetlands against development which threatens the existence of flocks of wild fowl. Each group starts from different, even mutually hostile, value postulates, but converges on the same policy outcome as a result of what they share, namely ecological (contextual) understanding.[5]

Norton adopts a policy-oriented, pragmatic approach to the discussion of environmental issues, according to which the actual outcomes are more important than the normative theorising which underpins them. However, there is a slight problem with this view if it is coupled with a 'broad church' approach which seeks to admit any value position 'undogmatically' provided it underpins agreed environmental policies to save the environment. For that would seem to allow the moralistic positions such as ecologism to be propounded with impunity. Yet, as we have seen, Norton regards these positions as something of a liability to the wider environmental movement.

The result is a curious hybrid position in Norton's discussion. He seeks to show that the moral theory which attributes intrinsic value to the non-human is open to philosophical objection, as being both internally incoherent and incompatible with the anti-dualist position which environmental philosophers have in general propounded. But he ends up by putting on his pragmatist's garb and in effect dismissing the discussion of such matters as something which will only be of interest to philosophers.

In fact, Norton's attempt to show the inadequacies of the moral theories distinctive of ecologism is not a philosophical side-issue. Certainly, ecologism will welcome convergence on its policy prescriptions by devotees of ideological and value-positions which rival its own. But these pragmatic considerations are insufficient to satisfy supporters of ecologism. For the latter believe that they have detected another strand of human arrogance which runs more deeply and is potentially more important than the one upon which Norton focuses his attention. This is, precisely, the idea that only human beings count morally. Ecologism takes itself to be expressing the truth of the matter in its view that, morally speaking, the non-human also counts.

Even if the acceptance of this leads in practice to no difference in specific policy prescriptions from those put forward by those with different moral views, there remains the claim that this is the correct view to take. While in present circumstances there may be convergence, there is no *a priori* reason to believe that this will always be the case. As Norton himself notes, alliances between different positions may break up as fundamental value differences come to the fore in new circumstances. Against Norton's own emphasis on science-based interconnectedness, for example, one might suggest that this will not help us to decide what we should do if we encounter a planet containing a biosphere with which we have absolutely no connection. May we exploit this without limit?

Political ideologists in general, for this kind of reason, do wish not simply for certain policies to be pursued, but for them to be pursued for what they regard as the right reasons, namely those which they support themselves. For only thus can it be guaranteed that the full project of the ideology will be consummated, not those bits of it which happen to overlap with those of rival ideologies, and which threaten to run off in entirely different directions.

Hence, the moral arguments of ecologism do have to be addressed. What, then, of Norton's criticisms of these arguments?[6] These focus on the claim that non-human nature is intrinsically valuable, that is, valuable in itself, irrespective of any instrumental value it may have for human beings. First, Norton convicts this claim of the incoherence of supposing that there can be value without a valuer. If the biosphere is intrinsically valuable, then this value will have to be posited as existing throughout the billions of years before there were any valuers around capable of recognising it. Such value has to be regarded as existing objectively in the world, and the realisation that it is there will have to be regarded as a discovery. This cuts off any connection between such value and human culture. Its presence can only be registered by an act of intuition, regarded as a direct registering of a quality without the relativisation of the property to any theory. This is because all theories (including scientific ones) are cultural constructs, yet *ex hypothesi* these values exist independently of culture.

Such acts of intuition in turn can only be understood in terms of a dualistic theory of human intelligence, which, following the Cartesian paradigm, supposes human intellects to have the capacity for the direct encounter with self-evident truths, for both are conceived of as inhabitants of a non-physical sphere. Norton suggests that, as well as resting on an epistemological basis which has been shown to be untenable, this dualist/intuitionist approach to the idea of objective value is a recipe for moral dogmatism. For you either 'discover' that there is such an objective value by exercise of your intuition, or you do not. Either way, no argument is possible.

However, there is a problem with the outright rejection of the idea of intrinsic value. This is that the concept is conjoined in a contrasting pair with that of 'instrumental value'. If all we had to operate with was the latter concept, then we would be committed to a vicious infinite regress. We could only ever explain the value of anything as a means to the attainment of something else, whose value in turn could only be explained in similar terms. At no point would we ever have succeeded in establishing the value of anything. For something to have value as a means to a given end only establishes the value of that thing if the given end is valuable. A means to a valueless end is itself valueless. In addition, of course, it has long been a staple of moral theorising that human beings are of value 'in themselves'. They are not to be treated solely as possessing instrumental value (which they undeniably do possess), but always also as possessing intrinsic value.

For such reasons, then, we appear to need to retain the concept of intrinsic value. How then do we do so without landing ourselves in the epistemological and metaphysical quagmire identified by Norton? His solution is to repeat the claim that there can be no value without a valuer and to use the term 'inherent value' to refer to the non-instrumental value which we need to be able to refer to if our value discourse is to be viable. Such a concept is, however, to be clearly seen to be a human cultural construct, not a given.[7] Once we grasp this point, then we can use the concept in the articulation of our value-positions without a commitment either to objectively existing properties which pre-exist, and continue after the possible demise of, human life and culture or to the dualist theory of mind which we need to posit to explain our culture-transcending discovery of them.

However, it is clear that Norton's resolution of the problem in effect clarifies the position of those who wish to attribute intrinsic value to the non-human, rather than refuting it. What is important for them is to use the privileging moral terms characteristic of humanistic moral discourse, however these are to be accounted for epistemologically, to characterise the non-human too. Ecologism may use the term preferred by Norton and speak of the 'inherent value' of the non-human; it may accept that such a concept is culturally constructed and deployed; it may as a result fully accept that the job of gaining acceptance of the claim cannot be a matter of simple dogmatic enunciation of it, as resting on an act of culture-transcending intuition. But the radically reorienting character of the ecologistic moral theory will remain untouched.

Further, the content of the claim will be much as Norton has characterised it, namely, that non-human nature has value in itself, irrespective of its instrumental value for human beings, and that it thus retains this value in the complete absence of human beings, past or future. A view analogous to this must, after all, be accepted by humanistic ethics with respect to human beings, if they too are granted inherent value. For example, a sole surviving member of the human species will have that value even though, *ex hypothesi*, the survivor can have no instrumental value for other humans and cannot have his or her inherent value endorsed by them.[8] This is a matter of the logic of the concept of inherent value, a concept which, as we have noted, is indispensable.

It might be said in reply that in such a case as the one just described the last human being will be there to attribute inherent value to himself or herself, which allows us to make sense of the situation and retain the principle 'no value without a valuer', whereas with respect to the hypothesis presented by ecologism, if there are no human beings or other relevant valuers around there can be no inherent value. However, this is an incorrect view. Suppose the last human being were racked with self-loathing. In his or her own eyes no inherent value may be discerned. Yet it is an implication of humanistic ethics that such a being has inherent value whatever that being's view of the matter. We do not normally accept the view that a human being who attaches no inherent value to himself or herself has the last word on the subject. We think there are good reasons for attributing such value to human beings whatever the view taken by individuals of themselves.

We should conclude, then, that in spite of Norton's arguments, the case for saying that non-human nature is intrinsically valuable has not been refuted. There is, of course, the great problem of how one is to establish such a position if one rules out, as seems desirable, the simple enunciation of self-evidence. We will shortly be examining Freya Mathews' argument which seeks to establish intrinsic value on the basis of her concept of 'self'.

Norton is correct in saying that there is an epistemological problem with attributions of intrinsic value to anything. But the same problem remains even if we regard such ideas as cultural constructs and use the term 'inherent value'. Norton does not explain how value-judgements in which the latter term is used are to be justified in terms of his own theory. What we need, if we reject the idea of 'objective values' waiting to be discovered, is an explanation of how such judgements attributing intrinsic/inherent value can come to have intersubjective validity.

In fact the problem with objective values is metaphysical, not epistemological. Roughly speaking, a value-judgement logically implies action – to attack or safeguard something. So an objective value would have to be discoverable rather than decided upon, and yet logically imply action on the part of the discoverer. This makes it a very odd kind of thing. Metaphysical oddness does not decisively refute the acceptability of the concept, for metaphysics and the more metaphysical areas of modern science, such as those explored by Freya Mathews, are replete with concepts we are hard put to make much sense of in terms of everyday thought. However, it is best to avoid such oddness if we can.[9]

What we may conclude from the consideration of Norton's discussion is, first, that ecologism is not committed to an objectivist position in ethics by its espousal of the intrinsic or inherent value of the non-human and, secondly, that the key issue which it raises does not concern the epistemological or metaphysical conundrums which have plagued moral theorising ever since it came into existence. Rather, the key issue is the first-order moral one of whether or not the non-human should be regarded as having any moral standing at all and, if it does, whether it is the same standing as we are used to attributing to human beings.

Let us now turn to consider the argumentative strategies which ecologism has tried to employ to gain acceptance for the moral considerability of the non-human.

Consistency Arguments

The obvious first strategy to employ here is to begin with an analysis of what feature accounts for the moral considerability of human beings and then to show that that feature is to be found in the non-human realm too. The argument at that point then turns into an appeal for consistency of treatment across the human and the non-human case. Resistance to such an appeal for consistency then can properly be criticised as a manifestation of arbitrary preference for one species over others – the charge of human chauvinism or speciesism which we have already encountered.

For the first strategy to work several points have to be clarified.

(1) We need, first of all, a correct, or at least defensible, statement of the basis of the moral considerability of human beings.

(2) We need to specify more precisely what the sphere of the non-human encompasses in this context: other individual creatures? Individual creatures from only some groups, such as those possessing sentience? Holistic entities, such as species and ecosystems?

(3) We need to vindicate the consistency argument, by showing that such consistency is a requirement of moral thought – some have questioned whether it is an appropriate consideration in such moral contexts.[10]

In this general area two versions of the consistency approach have sought to extend moral concern to the non-human. These approaches began from a different impetus and have been seen by some as mutually incompatible. However, it is more accurate to regard them as converging on to common ground.

The first version is that associated with the idea of 'animal liberation'. As exemplified in the work of Singer, Regan and others, this picks out features of human and non-human life such as the capacity to experience pleasure and pain, or to be a 'subject of a life' or to be autonomous in some sense.[11] It then argues that the possession by individual organisms of one or other of these capacities entitles them, whether human or not, to the concern of moral actors or, more strongly, to be accorded the right to certain kinds of treatment at the hands of such actors.

This version, as exemplified by Singer's utilitarianism, admits non-human creatures to the court of moral deliberation on an individual basis. It does not cover all non-human creatures, because the latter do not all possess the property which confers moral considerability. Typically they accord weight to other non-human entities, such as specific habitats, only in so far as these can be shown to have an instrumental importance to the well-being of the primary objects of moral concern.

The second version, such as that of Johnson,[12] focuses upon properties, such as the capacity for well-being, which prima facie can apply to holistic entities such as species and ecosystems as well as to individual organisms. Such theories thus extend moral considerability far beyond the focus upon individual (human) beings characteristic of anthropocentric ethics. This second approach is also characteristic of ecocentrists whose main focus of concern has precisely been the continued existence and well-being of holistic entities such as species, ecosystems (the 'land') and the biosphere as a whole.[13]

On some ecocentric views in this area (for example, those inspired by the Gaia hypothesis) one or other of these entities is itself viewed as a superorganism or quasi-organism, thus taking this version in the direction of the first version. The correctness of such views turns on the issue of the extent to which such entities as ecosystems possess self-organising powers, such as homeostasis, as well as on the issue of precisely how they are to be individuated.

The key idea of the second version, however, is that the capacity for self-organisation and its associated capacity for well-being are the crucial properties in virtue of which organisms such as human beings acquire moral considerability. Once it is accepted that such non-human entities do possess such capacities then it is possible to mount a consistency argument to the effect that according human beings moral considerability in virtue of their possession of such capacities commits us to attributing the same status to the non-human, including non-sentient organisms and the putative 'super-organisms' to which allusion has just been made.

Both versions, which are best conceived of as representing a series of positions on a single spectrum, are faced with moral dilemmas and moral trade-offs. The Singer–Regan approach does not take account directly of a whole range of life-forms which may, presumably, be sacrificed if necessary to safeguard the interests of the morally favoured classes (instrumental considerations permitting). The ecocentric approach appears to countenance the death and suffering of many sentient organisms for the sake of the well-being (integrity, beauty, and so on) of species, ecosystems and the biosphere. Both have the difficulty of dealing with conflicts of interest between the human and the non-human, although, as we noted above, a fuller, richer appreciation of what exactly is in the human interest is expected to obviate some of the apparent conflicts.

However, the differences between such theorists over precisely how to extend moral consideration to the non-human and how to strike the balance between the different moral considerations which their approach introduces, are not as important as the difference between them and devotees of a 'humans only' (or, to employ a more Kantian formulation, 'rational beings only') view of morality.[14] Clearly, too, what began as anthropocentric theories began heading in this direction long before the Green agenda became a subject of concern or debate, as is evidenced by the utilitarian approach of Bentham, who extended moral considerability to all creatures capable of feeling pleasure or pain. This, of course, is what one would expect of the consistency approach. Some theorists would be bound, sooner or later, to ask whether or not human beings could in consistency restrict their moral concerns to the human species alone, irrespective of any worries about the actual well-being of the non-human at human hands.

A key element in the consistency arguments is, of course, to resist the idea that the basis of human moral considerability is the possession by human beings of some property which is possessed by all and only members of the human species. Putative properties of this kind fall into two categories. First there are properties, such as the possession of an immortal soul, whose existence cannot be established by any known empirical means. Clearly, those who wish to resist the extension of moral considerability to the non-human are always able to cite such properties as the basis for the exclusive focus of moral concern on the human case. Of course, not everyone who holds this view also asserts such exclusiveness. The religious 'stewardship' traditions may allow a different route to the conclusion that the non-human may be the object of moral concern, albeit one which is still not empirically verifiable. However, the rights and wrongs of such views are theological issues, the resolution of which is notoriously difficult, particularly as the interpretation of texts, and not logic alone, is necessary for such resolution.

The second category covers properties, such as rationality, language-use, self-consciousness and so on whose presence is empirically ascertainable, deemed to be the basis of moral considerability and which are held to be exclusive to the human species. With respect to these, the strategy of ecologism has been to show that they are not characteristic of all human beings, or of any human being throughout their whole existence; and/or that they are not exclusive to the human species, so that at least some extension of moral concern beyond the human is necessary; and/or that they are complex properties which admit of degrees so that no clear moral demarcation-line can be drawn by their use; and/or that their relevance to moral considerability is open to question.

The conclusion to which these arguments lead is then that any property which is cited as relevant to moral considerability and possessed only by human beings turns out not to be possessed by all human beings, or by any human being all of the time; or, if it is possessed by all human beings all of the time turns out not to be possessed only by human beings. If we select the former properties, then we have to exclude from moral considerability all human beings some of the time, or some human beings all of the time. If we select the second set of properties, then the non-human is admitted to moral considerability. Since the former possibility is morally repugnant, we must select the latter, and admit the extension of moral considerability to the non-human.

There is, of course, one property which all human beings are guaranteed to possess and which only they possess, namely the property of being a human being. However, political ecologists argue, this must be a morally irrelevant property. Why should membership of a certain species in itself be sufficient to confer moral considerability on an organism? To think that it is sufficient is to be guilty of 'speciesism', the arbitrary moral preference of one species over others. A recent argument by Lynch and Wells has tried to show that such preference is not arbitrary and that being a human being is necessary and sufficient for moral considerability. Since this argument underpins the conclusion that there can be no issue of moral trade-offs between the human and the non-human, consideration of it will be deferred to the next chapter, which discusses such moral trade-offs directly.[15]

Consistency arguments are powerful tools of moral critique, and environmental philosophers who have employed them have thus far succeeded in showing that, try as they may, anthropocentric positions cannot prove solely by logic and empirically testable hypothesis that only human beings count morally. However, inevitably such arguments cannot be conclusive. It is always possible that some morally relevant property will be found which is possessed by all and only human beings and which resists the consistency approach. This is not too telling an objection. If one thing is clear from the history of human ethical thought it is that conclusive arguments are probably not to be found in any part of moral theory.

However, this point does make the second strategy an attractive one, which is to seek to argue for the intrinsic value of as much of the non-human world as possible. If we can establish such value by a plausible argument then the moral considerability of the non-human, given what 'intrinsic value' means, appears to follow without further argument. Let us now consider how such a case for intrinsic value may be made.

Intrinsic Value

An item has intrinsic value in so far as it is regarded as being valuable for its own sake. Mathews, as we noted in Chapter 2, argues that those systems which are self-preserving and self-realising – selves – possess intrinsic value. Let us consider her argument.[16] The key steps appear to be as follows:

(1) Any self, S, seeks to preserve itself and flourish
(2) therefore, necessarily S values itself for its own sake
(3) therefore, S necessarily has intrinsic value
(4) moral agents have the capacity to recognise the intrinsic value of S
(5) when they do so, they are committed, *ceteris paribus*, to preserving/protecting S.

This argument is intended to establish the intrinsic value of selves in the objective manner noted earlier in this chapter. That is, what conscious moral beings other than S do when they pass the judgement that S has intrinsic value is register a discovery they have made, not indicate a choice they have made to classify S as valuable for its own sake. The latter is possible, and Mathews, like Norton, labels it the phenomenon of 'inherent' value, to show that it is a matter of some valuer other than S choosing to confer intrinsic value on S.[17]

This argument also, however, keeps the connection between value and valuer which, as we have seen, many have claimed to be a logically necessary one. Thus, it promises to avoid the problems diagnosed by Norton. However, to get the result needed we have to interpret what is involved in 'valuing' in the right way. The valuer is, of course, S itself. Mathews tells us that on this view S does not have to possess self-awareness or even sentience to count as a valuer. Its intrinsic tendency to seek self-preservation and the conditions of its own flourishing is taken as an example of valuing.[18] Such a self, therefore, also has a 'good', specifiable once again in entirely objective terms.[19] That is, a human observer can, in principle, come to recognise the 'good' of the self in question – namely, what conduces to its preservation and flourishing.[20]

The key step in the above argument is clearly the move from (3) to (4). However, this is open to objection, as follows. What human observers can be brought to realise, if they accept the interpretation given in the previous paragraph of intrinsic value and 'valuing for selves', is that S has intrinsic value for S. This is what is implied by the 'no value without a valuer' principle.[21] But how does that establish that therefore S has intrinsic value for non-S, such as a human observer? Why should the intrinsic value of S for S have any significance for any non-S?

To avoid this problem, one might seek to eliminate step (2) entirely. The direct move from (1) to (3) would then amount to the attribution of intrinsic value to any self simply in virtue of its being a self and thus possessing the defining properties of selfhood. If we drop the presupposition that there is no value without a valuer, then a possible interpretation of this move is to see it as involving the registering of the existence of an objective property of S, supervenient on its other properties. However, such a value-property, as we noted above, has often been found to be metaphysically peculiar.

A more fundamental problem is that the claim that such an objective, supervenient property exists appears to involve a sheer assertion, the denial of which appears to involve no logical contradiction. This means in turn that we are starting to lose the logically compelling quality which Mathews' argument appeared initially to possess.

Further, if instead we reinstate the premise that there is no value without a valuer, the move from (1) to (3) will in effect register a human being's choice of how to value selves. It then is an instance of 'inherent' value, in Mathews' parlance. This avoids the 'peculiarity' problem, but leaves the argument open to the charge of subjectivism. That is, it may be claimed that there is no obvious argumentative strategy available to compel others to agree with your attribution of intrinsic value to 'selves'. We are free to make such valuations, but then they in effect register what liberal thinkers will quickly label 'lifestyle' choices, and the project of ecologism becomes nothing more than one view among many about the good life.

We should at this point take note that one kind of argument for compelling others to recognise the intrinsic value of selves seems clearly not to work. This involves the starting point that a human being is a self, and thus it may seem reasonable to press the claim that human beings (at least, normal, mature ones) are bound to attribute intrinsic value to themselves. One may then seek to argue that they must in consistency value other selves, including the simplest organisms.

The problem with this is that there is clearly no logical compulsion upon human beings who accept that they are intrinsically valuable themselves to recognise any intrinsic value in other selves, even other human beings. This is a version of the well-known problem in moral philosophy of how one moves from premises based on the first-person case (how you must view yourself) to conclusions based on the second- and third-person case (how you must view others). Kant solved this problem by restricting consideration to rational beings, establishing that rational beings necessarily value their own rationality, showing that rationality requires consistency, and then concluding that rational beings must, in so far as they value rationality, value other rational beings also.[22]

This solution works only because of a peculiarity of the concept of rationality. It does not apply to non-rational beings, and so does not serve the cause of ecologism. But the idea of a self, as characterised by Mathews, seems to have no similar feature to ensure that human valuers, although selves, are compelled, on pain of self-contradiction, to grant intrinsic value to selves other than themselves. It is true that Mathews, and others, have wished to claim that non-human selves are, as self-preserving and self-realising entities, in the same category of 'ends-in-themselves' as Kant puts rational beings.[23] But Kant clearly views rationality as an essential feature for entry into that category. There appears to be no reason to convict him of a logical error in doing this, and to convict him of moral shortsightedness appears to be purely question-begging.

A further argumentative strategy ingeniously offered by John

O'Neill, is in effect to offer points (1), (2) and (3) of Mathews' argument above, and then to suggest the following. Assume that moral agents accept that other selves, which are not moral agents, have a good, and that such agents should act to further those other selves' good. This amounts to those agents' ascent to a higher, universal standpoint from which the goods of other possessors of intrinsic value, whether moral agents or not, are placed on a par with their own. The adoption of such an objective, universal standpoint, a capacity which is peculiar to moral agents such as ourselves, is itself a crucial part of the good for such beings. Just as in cases of human friendship we find, as Aristotle argued, that caring for another for their sake is part of our good, so also is taking account of the good of other selves.[24]

This argument, of course, may be seen as a species of argument from human self-interest, and thus may be viewed as unacceptably anthropocentric from the point of view of ecologism. However, as we have seen, it is very difficult to avoid some form of anthropocentrism, for the same reason that within purely human-centred moral theory in general it is very hard to answer the sceptical question 'why should I be moral?' without saying in some way that it is in your own best interests to be so. This may be viewed as a species of egoism, the apparent opposite of what morality, with its commitment to altruism, is supposed to uphold.

As with moral debate in general, the real problem with this reply of O'Neill's is to convince sceptics that the acceptance of the good of others as a principle of one's own action is indeed part of one's own good, properly understood. When it is recast as a prudential argument, particularly as applied to human beings alone (for human beings can sanction each other), the argument becomes more readily acceptable to the sceptical mind. But neither moral theorists in general, nor supporters of ecologism in particular, can accept that as an adequate strategy. As a moral argument, however, it is clearly not a knock-down one. Someone who rejects it commits no intellectual misdemeanour. Those who do accept the argument will, however, firmly expect the life of the rejector to go less well than it otherwise would.

We may conclude, then, that Mathews does not succeed in establishing the intrinsic value of selves which are not persons, and thus not moral agents, in a way which compels acceptance of this value by other selves which are moral agents. But ecologism still needs some such argument.

Intrinsic Value and 'Wonderfulness'

Let us now consider another possible argument for intrinsic value of the non-human. We may begin by noting that what many people feel about other organisms and species is somewhat as follows. Each different creature – animal, plant, bacterium or whatever – is in its own way something wonderful. Each represents a different manifestation of the larger wonder of life. One cannot regard life as something wonderful and yet be indifferent to its concrete manifestations, for life only exists in its concrete manifestations. We have here returned to an aspect of that phenomenon referred to in Chapter 3 as 'biophilia'.

Some manifestations reveal their wonderfulness to the human eye more readily than others. However, one of the rewards and benefits of disinterested study of creatures, however humble or 'primitive' or small, is the revelation it gives of the particular way in which they embody their own portion of the wonderfulness of life.[25] Those who popularise and disseminate the findings of biological science perform the vital task of disseminating this revelation to millions who have not the time or expertise to acquire it for themselves. Of course, some people will always be blind or indifferent to such things. But many, perhaps most, people have at least some capacity to recognise this wonderfulness.

The important point here is that this quality of being wonderful is attributed by people to the things themselves. It is not regarded as a purely subjective reaction.[26] People who have this view then naturally and reasonably also take the view that the potential loss of each separate portion of wonderfulness should be avoided if at all possible, particularly when it results from an avoidable human action.

Of course, wonderfulness is not the same as pleasantness or harmlessness. There is much that strikes even the most avid explorer of the life-forms of our planet as ghastly, and some of it is directly harmful to human beings, such as various bacteria and viruses. But even entities such as these are manifestations of wonderfulness.

In addition, as far as we know, we are the only creatures capable of recognising the wonderfulness in question. But that is irrelevant to the correctness of the claim that the myriad forms of life are wonderful and that this constitutes a reason – the strongest reason there is – for moral actors to accord moral considerability to such life-forms.

Further, once we realise that individual specimens are only the temporary embodiments of this wonder – for all individuals are mortal – and that individuals only exist in the form they do because of the existence of species of which they are members, then we have a direct argument for the preservation of species, and of the habitats within which they have evolved and to which they owe their specific form. That is, we have a reason for preserving species in their habitat, not for preserving simply the genetic material from which the species may be constructed.

Is this argument a moral one at all? The concept of wonderfulness here appealed to may appear to be an aesthetic rather than a moral concept. If that is how the concept should be categorised then this may tend to dowgrade the argument in the eyes of many, for an aesthetic property, however attractive, may seem too trivial a basis upon which to found a claim of moral considerability.

However, if the characteristic appealed to by this argument is an aesthetic one, that does not inevitably detract from its force as a moral consideration. There clearly are moral concepts whose nature is as much aesthetic as moral. Moral goodness is itself a kind of beauty, moral badness a kind of ugliness. If we are trying, in a moral context, to elucidate the concept of intrinsic value, then the idea of wonderfulness, elusive as it may be, looks to be an excellent way of doing it. Even here there is an instrumental consideration relevant to human beings. For things which are wonderful are among the best candidates to be sources of human joy.

It might be urged in criticism of this argument that it fails to establish its conclusion because being wonderful is at most a necessary, not a sufficient condition for being properly regarded as intrinsically valuable. We may, after all, find an intricate machine to be wonderful, but we usually regard all machines as being of purely instrumental value. We certainly do not say that such items possess moral considerability.

This point may be met by saying that to the extent that an artefact possesses the quality of wonderfulness it does possess intrinsic value and its continued existence does have a moral claim on us. However, machines, unlike organic beings, may perfectly well exist indefinitely as single specimens. It is even possible for them to remain accessible purely as blueprints. Thus, we should, morally speaking, preserve specimens or specifications of such wonderful artefacts just because they possess intrinsic value. In the case of other artefacts, such as paintings, their uniqueness requires that we maintain in existence the actual specimens. They cannot be reconstituted from blueprints as can machines. In some cases, such as wonderful buildings, there is a requirement akin to that which applies to species, that we seek to preserve the 'habitat' or context within which its full wonderfulness becomes apparent.

However, does it make sense to say that the moral wrong done in the wanton destruction of a wonderful artefact is a wrong done to the artefact? Is not the moral responsibility held towards human beings with respect to the preservation of the wonderful artefacts?[27] There is force to this point. It is at this juncture, however, that we can bring in those properties which were adduced in the 'consistency' arguments and in Mathews' argument for intrinsic value. We can use them to delimit those entities within the category of wonderfulness which possess moral considerability in virtue of their intrinsic value. Those wonderful entities which are, in Mathews' formulation, selves, being self-organising, self-preserving and possessing a good of their own, are intrinsically valuable and morally considerable. Plainly, human artefacts (at least until something like Frankenstein's monster becomes a reality) do not fall into this category.[28]

Is not all the work in this argument being done by the elements introduced under the category of 'self'? Does 'wonderfulness' add anything of importance? It does add the element which is necessary to establish the intrinsic value of the organisms for moral agents, rather than for the organisms themselves. The recognition of the wonderfulness of such selves gives us the reason for attributing to them intrinsic value which was missing in Mathews' argument. This is needed in turn as a necessary condition for the moral considerability of such entities, once the other necessary condition, of 'selfhood', is also recognised.

However, it may well be asked whether the concept of being wonderful has enough descriptive content to enable it to be anchored in some objective way in the entities to which it is applied? Or is it purely a value term, meaning something like 'to be regarded as intrinsically or instrumentally valuable'? Clearly, if it were solely the latter then it would be useless for the purpose of justifying the claim that something possessed intrinsic value.

There is no doubt that the word may be used as a vague expression of positive evaluation. But it is also clear that in its considered uses there are features of the item characterised as 'wonderful' which have to be indicated in order to justify the application of the term. These features do not comprise a finite set to be found in all cases. But it is clear that such properties as complexity, intricacy, fittingness for a given purpose, beauty, elegance and economy of means, when present to a sufficient degree and in combination, are appropriate bases upon which to justify the application of the term to an entity. Hence it is a sufficiently 'thick' value term to play a justificatory role in discussions of value.[29] To put the point in a different way, not all things which possess intrinsic value do so because they are wonderful in the sense just outlined.

This, then, is the argument for claiming that non-human life-forms possess intrinsic value and that they are therefore morally considerable. To recollect the point behind this argument, the various ways in which non-human organisms and entities possess instrumental value for human beings are at present undeniable and important. However, the reliance on instrumental value alone to justify the preservation of such things is problematic, for such value is contingent and limited in scope.

The consistency arguments, while valuably extending moral considerability on the basis of the comparison with the human case, are inevitably always open to the possibility of refutation on the basis of the refinement of purely anthropocentric positions. Thus, from the point of view of ecologism, some argument for directly attributing intrinsic value to the non-human is required.[30] As we saw when discussing the arguments of Norton, this concept need not lead us into a metaphysical and epistemological quagmire if handled properly. The argument from wonderfulness aims to secure interpersonal validity for judgements of intrinsic value by referring to properties in the items so judged which justify the judgements. Such judgements are not independent of theory, for the properties in question require for their detection theoretical understanding of different degrees of sophistication. Thus, the judgements are not mere assertions of intuition of the kind criticised by Norton. But, as with all value-judgements, they may be rejected without logical impropriety.

What ecologism can show in this area, therefore, does not compel acceptance on pain of self-contradiction. But in this it is no worse off than pure anthropocentric theories which also have to assert intrinsic value of human beings (or, more generally, persons). Ecologists, thus, have no reason to abandon their attributions of intrinsic value to the non-human even if they do not possess a logically compelling case for such attributions.[31]

Notes

1. See Naess, *Ecology, Community and Lifestyle* (1989), Devall and Sessions, *Deep Ecology* (1985) and Fox, *Toward a Transpersonal Ecology* (1990).
2. Tim Hayward, 'Anthropocentrism: a misunderstood problem' (1997), p. 61.
3. The discussion in the next section focuses on Bryan Norton, *Toward Unity among Environmentalists* (1991).
4. As he says on p. 237 of *Toward Unity among Environmentalists*, 'If the target is arrogance, a scientifically informed contextualism that sees us as one animal species existing derivatively, even parasitically, as a part of a larger, awesomely wonderful whole should cut us down to size'.
5. Norton 1991, p. 202.
6. Norton 1991, pp. 234–7.
7. Norton, p. 236.
8. As John Barry has reminded me, John O'Neill, in chapter 3 of his *Ecology, Policy and Politics* (1993), has an argument for saying that the dead can be harmed by the living, by means of the latter's failure properly to deal with the reputation and uncompleted projects of the dead. Hence, a sole surviving member of the human species might have an instrumental value for the dead. I doubt it, however. His or her instrumental value rests on his or her passing of something on to *his or her* posterity, and that is ruled out in this case. We will return to this argument of O'Neill's more directly in the next chapter.
9. See J. L. Mackie, *Ethics: Inventing Right and Wrong* (1977), pp. 38–42 for further discussion of the metaphysical 'queerness' of objective values.
10. See, for example, David Cooper, 'Other species and moral reason' (1995).
11. See Tom Regan, *The Case for Animal Rights* (1983) and Peter Singer, *Animal Liberation* (1995).
12. Lawrence Johnson, *A Morally Deep World* (1993).
13. See, for example, J. Callicott, *In Defense of the Land Ethic* (1989).
14. The *locus classicus* for the exploration of these differences is K. Goodpaster, 'On being morally considerable' (1978).
15. Tony Lynch and David Wells, 'Non-anthropocentrism? A killing objection' (1998).
16. The following presentation of Mathews' argument is my attempt to encapsulate the ideas presented on pp. 118–19 of *The Ecological Self* (1991).
17. Mathews 1991, p. 178.
18. Mathews 1991, p. 104.
19. Mathews 1991, p. 103.
20. A similar account of the moral considerability of non-human organ-

isms is that given by Paul Taylor in *Respect for Nature* (1986), p. 75, who argues that we must attribute 'inherent worth' to entities which have a good of their own, irrespective of any intrinsic or instrumental value attributed to them by human beings and without reference to the good of any other being. Taylor argues that the proposition that some entity has inherent worth of this kind entails that that entity is morally considerable and that all moral actors have thus a prima facie duty to preserve and promote its good as an end in itself and for the sake of the entity whose good it is.

This enables us to distinguish the moral status of organisms from that of non-living complex entities with interactive parts, such as machines, for the latter do *not* have a good of their own.

21. This principle is endorsed by Mathews 1991, on p. 105.
22. Immanuel Kant, *Groundwork of the Metaphysic of Morals*, trans. by H. Parton as *The Moral Law* (1948), p. 91.
23. Mathews 1991, p. 119.
24. See John O'Neill 1993, pp. 23–5.
25. It is appropriate to refer at this point to the theme of Rolston's already encountered in connection with our earlier discussion of meaningfulness (Chapter 2), namely the role of narrative. For our sense of wonderfulness is embodied in the story we tell of life-forms – their 'natural history'. It is not simply a matter of enumerating their parts and their interconnections.
26. H. Rolston, *Environmental Ethics* (1988), p. 26, justifies this feeling in the following way:

> Humans desire an environment sophisticated enough to match their wonderful brains. From another perspective, we ask whether such wonder (taking place in wonderful brains) can be generated except in the presence of something worthy enough to induce it, which suggests that nature is intrinsically a wonderland.

This other perspective is, of course, that of naturalism, which sees human beings as interconnected with the rest of nature. This then makes sense of the particular valuings of human beings as what is to be expected of creatures emerging in the course of the value-creating processes of the natural world.

27. This argument differs from the position put forward by Attfield which accords intrinsic value to species and individual organisms, but refers to artefacts as possessing 'inherent' value – see R. Attfield, *The Ethics of Environmental Concern* (1991), p. 152.
28. Nor do natural but non-living entities such as mountains, rocks and rivers. Rolston 1988, p. 199, argues that although such things lack the biological capacity for self-maintenance and organisation, they can be individuated and are 'the products of systemic nature' just as life-forms are. They have their own careers embodying the natural

processes which gave rise to them in the first place. As such, they possess intrinsic value and even moral considerability.

It is certainly plausible to suggest that, within the perspective of ethical naturalism, they are wonderful, worthy of respect, and should be treated with due appreciation of their status as part of the ongoing process of nature. But this in itself does not seem to amount to moral considerability, for which the idea of 'suffering harm' (as opposed to damage) and 'possession of a good of their own' are essential.

29. The distinction between 'thick' and 'thin' value terms is explained in Bernard Williams, *Ethics and the Limits of Philosophy* (1985), chapter 8.
30. Rolston 1988, pp. 217–18, makes the important point that to say that something possesses intrinsic value is not to imply that it is being considered in a way which isolates it from other things: 'Intrinsic value is a part in a whole, not to be fragmented by valuing it in isolation' (p. 217). He then suggests that intrinsic value 'becomes problematic in a holistic web'(p. 217).

    However, the fact that intrinsic value is always attributed to things in their context does not make it problematic. It may be difficult to individuate the objects of the judgement of intrinsic value, and very often they also possess instrumental value in virtue of their interconnectedness with other things. But that does not show that the concept of intrinsic value makes no clear sense, or that it is indistinguishable from the concept of instrumental value. It is rather that the context of things-in-a-context is needed to reveal the character, and thus intrinsic value, of those things.
31. Onora O'Neill offers environmental philosophers an obligation-based ethic which she believes avoids these problems with value-judgements – see O'Neill, 'Environmental values, anthropocentrism and speciesism' (1997). Discussion of her interesting ideas would take us too far away from the topic of this book. I discuss her ideas in my 'Environmental ethics – values or obligations?' (1999).

5

Moral Considerability and Moral Trade-offs

ECOLOGISM'S MORAL POSITION BEGINS, then, with a distinctive ultimate value postulate, namely that 'all life-forms are intrinsically valuable'. From this, ecologism concludes that all life-forms are worthy of moral consideration. That is, moral agents are required to consider the interests and needs of such life-forms when deliberating over what course of action they ought, morally speaking, to follow.

Crucially, however, the property possessed, intrinsic value, is intended by ecologism to admit of degrees. Some life-forms have greater wonderfulness, and thus intrinsic value and moral considerability, than others. In view of this, one might suppose that there is therefore a clear contrast between ecologism's basic moral postulate and that which underlies traditional anthropocentric theories – 'each person matters equally', to use Kymlicka's formulation.[1] This is misleading, however, for personhood also admits of degree. It is thus an issue for purely anthropocentric theories whether individual human beings who are not fully persons – foetuses in the womb, new-born babes, those who are severely brain-damaged, demented, irreversably comatose, and so on – do matter equally. This is an issue for ecologism too, and is as subject to debate within ecologism as it is within traditional moral and political philosophy.

This leads on to a key point about the logic of ecologism as a political philosophy. It operates with fundamental moral concepts which are able to encompass both humans and non-humans. It then locates moral and political philosophy as it applies to humans as a sub-section of ecological moral and political philosophy. The characteristic features of this sub-section derive from the species-specific characteristics which give human beings a distinct moral status, albeit one which is related to that of other life-forms. But the key point is that ecologism tells us that there are certain claims which all life-forms have upon moral agents, that we are all inhabitants of the same moral space.

Ecologism is thus more complex than many traditional moral theories in attempting to encompass, from the start and as a matter of principle, all life-forms within its moral thinking. It is also more complex in another direction. Its emphasis upon the interconnectedness of life-forms leads it to spread moral concern wherever the connections lead. It does not draw what it sees as arbitrary boundaries around species, aspects of the biosphere of the planet or human groupings. Thus, it also seeks to encompass the moral claims upon existing people of human contemporaries living outside their own societies (contemporary aliens) and of future generations of people and non-human life-forms.

Is this too ambitious a goal? Does ecologism's attempt to do justice to what it sees as the moral claims emanating from directions barely noticed in traditional ethics – the non-human, aliens and future generations – lead to a mass of conflicting and often hypothetical moral claims which it is impossible to reconcile even in one mind, let alone in a public domain which is the only arena within which such moral claims can effectively be met?

In response to this concern, we will in this chapter attempt to search for some principles to order this complex set of considerations. We will require a principled way to think of the degrees to which different forms of the non-human may be supposed to have claims on (human) moral agents. This concerns how to conduct moral trade-offs between the interests of beings each of which possesses moral considerability, but to different degrees. Such principles will form a central part of what has been dubbed by Low and Gleeson 'ecological justice', or justice between human beings and the rest of the natural world.[2]

This is to be contrasted with the issue of social justice between human beings, which must now, ecologism maintains, encompass contemporary aliens and future generations. It must also embody principles of what Low and Gleeson dub 'environmental justice', which concerns the distribution of environmental benefits and burdens among human beings.[3] In this chapter we will primarily concentrate upon the extension of considerations of social, including environmental, justice to future generations.

However, we first need to address an argument from within the Green camp, presented by Freya Mathews, which claims that intrinsic value should be attributed to all life-forms to an equal degree. It is important to try to show that this argument fails, for a workable theory of moral trade-offs between beings possessing intrinsic value requires that different degrees of intrinsic value be attributed to them.

Egalitarianism, Complexity and Intrinsic Value

It will be recalled that Mathews offers a theory of the intrinsic value of what she calls 'selves' – systems which possess the features of self-maintenance and self-realisation.[4] On her account this encompasses all organisms, whether sentient or not, and some non-organic entities too – ecosystems, the Earth's biosphere, the cosmic self. From this position she extracts the further claim that, prima facie, different selves possess different degrees of intrinsic value. They possess different degrees of intrinsic value in so far as they possess to different degrees the capacity to protect and realise themselves. This means, in effect, possessing differentially the ability to cope with environmental threats to their existence and flourishing. In general, the greater degree of complexity a self possesses the greater its ability to be self-sustaining in varied and changing environments, and the greater the degree of intrinsic value it will possess.[5] As human beings are clearly the most complex organisms of which we are aware on this planet, it follows that human beings possess the highest degree of intrinsic value of all organisms on this planet.[6]

However, this position represents only the first phase of the argument concerning equality of intrinsic value. It rests, Mathews claims, on viewing selves in isolation, abstracted from the ecosystems in which they are embedded. When, however, we consider them in actual, complex ecosystems, outside of which they cannot in fact survive, however complex they are, then we immediately re-encounter the basic fact of interconnectedness. To take the example she gives, a Blue Whale is enormously more complex than the krill upon which it exclusively feeds.[7] But without the krill the Blue Whale cannot survive, and probably without the Blue Whale, to regulate its numbers, the krill will not survive (or at least flourish) either. A Blue Whale and a specimen of krill, considered in isolation, possess very different degrees of intrinsic value. But when Blue

Whales and krill are considered in their full interrelatedness, an equalisation of intrinsic value ensues. Since intrinsic value follows ability to be self-protecting and self-developing, the krill participates, via interconnectedness, in the Blue Whale's ability to survive, and thus acquires some of that higher intrinsic value. In so far as the Blue Whale relies on the krill for its survival and flourishing, it loses some of its intrinsic value to the krill.

Considered, therefore, as species, Blue Whales and krill are on a par with each other. However, individual specimens of Blue Whales remain more intrinsically valuable than individual specimens of krill, provided we prescind from considerations of scarcity, when the 'principle of maintaining biodiversity' comes into play to increase the intrinsic value of a rare, but less complex, self.[8] Hence, if we have to choose between killing a Blue Whale and killing a specimen of krill, then the Blue Whale is correctly chosen.

Hence, the fact of ecological interconnectedness equalises, at least at the level of species, the intrinsic value which selves would otherwise possess to different degrees. This, of course, thereby equalises what would otherwise be differing degrees of moral considerability of different selves.

There are two elements in this analysis which are valuable for the development of the ethics of ecologism. The first is the connection between degree of complexity and degree of intrinsic value. Although she partially limits the scope of this connection by the ecological interconnectedness argument just given, Mathews provides the basis here for some method of determining moral trade-offs between the conflicting interests of different selves. The second valuable idea is that the basic degree of intrinsic value of a self may be increased by the fact of scarcity,[9] on the assumption that the maintenance of biodiversity is morally required. This too is an indispensable step for the development of a viable theory of moral trade-offs for ecologism. Both of these elements will be returned to below.

However, it will be necessary to detach the intrinsic value of a self from its capacity for self-preservation and self-development and attach it directly to its degree of complexity. This is partly because it is very unclear why such a capacity has anything other than instrumental value, even for the self in question. It is also because the use which Mathews makes of the connection between the capacity and intrinsic value in her 'ecological interconnectedness' argument is open to serious objection.

An analogy may help to clarify this objection. Consider the relation between two items – not this time selves – which possess different degrees of aesthetic value, namely aesthetically successful oil-paintings (that is, not mere daubs) and the oil-paints out of which they are made. The oil-paintings clearly depend for their existence upon the paints – no oil-paints, no oil-paintings. The paints probably depend for their existence on the paintings – there is no call for oil-paints if no oil-paintings are to be produced. Further, the paints and the paintings are mutually determining in the best traditions of holistic interconnectedness. That is, to produce paintings with certain specific characteristics, paints of a certain kind are called for. From the other direction, oil-paints of that kind inevitably give the paintings in which they are used a certain specific character.

In this example, oil-paintings and oil-paints are intended to be equivalent to the Blue Whales and krill of Mathews' example. The conclusion to which the argument is pointing, however, is not the equivalent one to hers. It clearly does not follow from the interconnections of existence and character just outlined between the oil-paint and the oil-paintings that the two become equal in aesthetic value. My accumulation of tubes of high-quality, aesthetically pleasing oil-paint will not give me a cheap way of acquiring a Rembrandt. The contribution such paint makes to the existence and aesthetic success of the painting, or to the capacity of the painter to create an aesthetically pleasing painting, does not mean that the paint itself acquires some of the aesthetic value of the painting.

Similarly, the facts of holistic interconnectedness cited by Mathews do not validate a thesis of equality of intrinsic value of the selves so interconnected. This is just as well, for in conflicts of value in which such modifying factors such as scarcity and the preservation of biodiversity do not operate, the thesis of equality of intrinsic value makes every decision a moral 'toss-up' (which is how she views the choice between preserving Blue Whales as a species and preserving krill as a species). There may well be such toss-ups, but a theory which makes most moral decisions arbitrary, as the egalitarian position apparently does, is simply not credible – or morally acceptable, either.

Ecological interconnectedness is clearly not devoid of moral significance. It clearly gives organisms added instrumental value for each other, which is often a matter of moral importance. It also possibly creates non-organic selves, such as ecosystems, which are candidates for the possession of intrinsic value. But it does not appear to substantiate the thesis of equality of intrinsic value suggested by Mathews.

Complexity and Degrees of Moral Considerability

We have concluded, then, that although all life-forms are morally considerable, this is a characteristic which admits of degrees. In order to elucidate this further, we may begin with the point that it is possible to arrange life-forms along a spectrum of complexity which essentially is determined by the evolutionary history of life on this planet. This is not to say that it is the point or purpose of evolution to produce increasing complexity. We are simply noting that, as a matter of fact, increasing complexity has occurred.

Thus, at one extreme are the life-forms on the edge of the living/non-living divide, essentially strands of DNA, such as viruses. Their chief characteristic is that of self-replication. Allowing for mutations, specimens of these are largely interchangeable. That is, the notion of individuality is largely absent from these life-forms. We then move to single-celled organisms; to complex organisms which start to show greater signs of individuality; to complex organisms with much greater individuality, with this emerging quality being bound up with such qualities as consciousness, self-awareness, capacity for complex behaviours and so on. Finally we reach at the other end of the spectrum human beings, possessing the highest degree of individuality of which we are presently aware, bound up with personhood – full self-consciousness, capacity for autonomy and foresight and a moral sense.

The degree of moral considerability becomes higher as we move along this spectrum from simpler to more complex life-forms. If, as argued in the last chapter, wonderfulness is a thick concept, with a descriptive content to be spelled out in terms of intricacy, fittingness for a given purpose, beauty, elegance and economy of means, then as these increase with complexity so must wonderfulness. As intrinsic value is based upon wonderfulness, and as moral considerability is based upon intrinsic value, then the intelligibility of the increase of moral considerability with greater complexity becomes readily apparent. Further, as we move along this spectrum, the kinds and range of benefits and harms which life-forms can undergo and experience, their capacity for suffering, their degree of interchangeability and other properties change, also becoming more complex.

However, all life-forms count morally, so that when the interests of a given life-form clash with those of the most complex life-forms, namely the moral agents, then the latter are morally required to seek to respect those interests where possible. If that is not possible, and a choice has to be made of which interests to sacrifice, then the basic rule is that the interests of the more morally considerable must be given priority. However, the scope for action of moral agents is limited by what is needed to take account of the moral considerability of other life-forms.

A problem which arises for ecologism at this point is, obviously, that if this is all there is to be said on the matter, then prima facie ecologism's moral prescriptions will not materially differ from those of purely humanist moralities. It looks as though, on the above rule, human interests will always triumph over the interests of the non-human, for, as already admitted, human beings have the highest degree of moral considerability of all the life-forms we know.

To avoid this outcome, we need to introduce some further distinctions within the concept of moral considerability. First, when life-forms are of low degrees of individuality, that is when individual specimens are interchangeable with respect to characteristic patterns of behaviour, then the prime object of moral concern, that which is morally considerable, is the species. With life-forms showing high degrees of individuality the prime objects of moral concern are the individual specimens, although even here the species is not devoid of moral considerability, for the reasons given in the last chapter in the course of the discussion of the implications of the wonderfulness of species.

The degree of intrinsic value which a low-individuality species has is much greater than the degree of intrinsic value of any given specimen of it. The intrinsic value of the latter, therefore, can usually be overridden without difficulty when clashes of interests arise with the interests of entities possessing higher degrees of intrinsic value. It is possible in such cases for the intrinsic value of individual specimens to be so low (even though not zero), that the instrumental value of such specimens for creatures with greater intrinsic value will easily outweigh it. The use of bacteria in medical experiments would be a clear case in point – although there will be cases which are much harder to decide.

However, if such species should start to decline, the specimens will increase in their degree of intrinsic value until their intrinsic value rivals that of the species itself. This is intelligible, for with declining numbers the species and the individual specimens start to coincide to an increasing degree.

This situation of decline brings into view another dimension of evaluation – a dimension noted by, for example, Dale Jamieson.[10] This is the dimension of urgency. The intrinsic value of something the existence of which is under threat, becomes the basis for demanding urgent action to preserve it and rescue it from the threat. This urgent demand is directed towards those moral agents in a position to perform the action.

This means in turn that situations may arise in which specimens which normally have low degrees of intrinsic value increase their intrinsic value and concomitantly the degree of urgency of the preservative actions required of moral agents also increases. When this involves a clash with the interests of specimens (such as human beings) which are inherently of a higher degree of intrinsic value, then the degree of urgency of the action required to preserve the threatened species/specimens may justifiably lead to those interests being overridden.

Hence, human interests will not automatically trump the interests of beings with lower degrees of intrinsic value and moral considerability. Urgency may transform the moral claims of the latter beings. Indeed, at this point we begin to encounter the idea of interspecies, or ecological, justice. For we can formulate the point just made by saying that it is a contravention of justice for human moral agents to allow the non-basic interests of human beings to trump the basic interests of non-humans. Such a contravention is unjust because it fails to give the non-human its due – which is the requirement that its moral considerability be recognised. As noted above, however, we need to give an account of how such ecological justice relates to social justice between humans.

A further complication to note is that it is possible that whole species, especially at the simple end of the spectrum, may themselves show such little difference from other species of the same genus, that they are almost devoid of individuality, in which case they will have lower degrees of moral considerability than would otherwise be the case. This is because their disappearance will leave other species in the same genus which will be bearers of the moral considerability of the species which has disappeared. In such a situation, the threat to the species concerned may not lead to a concomitant rise in the urgency of the action required to preserve it from extinction.

In addition, some species may be rightfully exterminated by moral agents, notwithstanding their moral considerability, if their *modus vivendi* causes great harm (such as suffering and death) to creatures with high degrees of moral considerability and they possess no redeeming instrumental value for such creatures. The smallpox virus is a candidate example. Even here, however, the moral agents to whom issues of moral considerability are addressed are required to ascertain that the harmful *modus vivendi* is unavoidable, and only proceed to exterminate where no alternative means of maintaining that life-form in existence is possible (within present knowledge).

When clashes of interest occur between life-forms, neither of which is a moral agent, as in the case of predators and prey, there is no automatic requirement for moral agents to intervene to referee the clash when they become aware of it. Moral agents are bound not to be indifferent to the commitment of moral wrongs, and are obliged to take steps to prevent them from occurring. However, life-forms which are not moral agents logically cannot commit moral wrongs. In virtue of their moral considerability they have moral claims against those life-forms which are moral agents, but not against life-forms which are not.

The intervention of moral agents in such circumstances is only prima facie required to preserve the existence of species. However, there are limits to what moral agents can be expected to attempt. For example, if a species is nearing extinction as a result of the workings of overwhelming natural forces, such as climate change, then no action is really possible which is likely to help. A further limit stems from the fact that one key component in the moral considerability of life-forms is that they be allowed to live lives of the kind typical of their species, within circumstances comparable to those in which they evolved and reached their distinctive form, with all the vicissitudes involved in that, including predation. We should not, of course, count within those vicissitudes the morally objectionable activities of moral agents. The latter are subject to moral criticism and restraint.

As we will be discovering more fully when we later turn to consider the economic implications of ecologism, the moral considerability foundation of ecologism rules out the idea that decisions about trade-offs between the interests of moral agents and nonmoral life-forms can properly be conducted purely in terms of cost-benefit analysis. The issue of making such trade-offs is a moral one, not simply an economic one to do with the efficient use of resources. Non-human beings are an economic resource for humans, as they are for each other, so that cost-benefit analysis has its place in the formation of the necessary information upon which a decision is to take place. But, as Booth and others have argued,[11] cost-benefit analysis cannot settle the matter on its own, notwithstanding the attempts of supporters of cost-benefit analysis to cover moral dimensions by attempting to price such desiderata as existence-value and option-value.

When the interests of human beings are subordinated to the interests of the non-human, then the human beings whose interests are overridden may be entitled to some compensation, depending on the degree of importance which the interests overridden have for the living of a reasonable or decent human life. This is one reason among others why social policy among human beings is a matter of concern for political ecologists, as Booth has pointed out.[12]

However, before leaving this discussion of moral trade-offs there are two further issues which need to be addressed. The first is the claim, by Lynch and Wells, that the effort to work out a principled method of reaching moral trade-offs between all morally considerable beings, human and non-human, is misconceived, because in any clash of interests between the human and the non-human, human beings must, morally speaking, always be given priority simply in virtue of being human.

The second is an issue of justice which cannot be deferred to later. This concerns ecologism's commitment to taking into account the interests of future generations in moral deliberation with respect to our relations to the non-human world.

Do Human Beings Always Count More?

Lynch and Wells seek to defend a 'weak' form of human chauvinism, characterised by the Routleys as 'the invariable allocation of greater value or preference, on the basis of species, to humans, while not however entirely excluding non-humans from moral consideration and claims'.[13] They accept that human beings have moral obligations to other animals, as in the avoidance of the infliction of needless suffering upon them.

However, their key claim is that in situations where we have to choose between a basic interest of a human being, such as the preservation of its life, and an equally basic interest of a non-human, then we human beings are morally required to give the priority to the human being simply because it is a human being. To take their example, upon discovering a human being under attack and threatened with death by a non-human animal, even a rare and noble one, we should, morally speaking, kill the animal to save the human without hesitation or reflection, just as soon as we recognise that it is a human being involved. This may lead to regret on our part, if we find, for example, that we have had to kill a specimen of a rare and beautiful species of predator, a regret which may be heightened if we also discover that we have thereby saved the life of a moral monster, such as an Adolf Hitler. But however we specify the situation so as to envisage an admirable non-human versus a non-admirable human, the fact remains that in such a situation the human, simply in virtue of being human, always has the prior moral claim.

What Lynch and Wells are most concerned to resist is the placing of human beings and non-human beings upon a single scale of moral calculation. They argue that even if, in an example involving a clash of basic interests, a non-human should in fact possess some attributes, such as intelligence, a rich emotional life and so on, to a greater degree than a human being, such as someone with a severe mental handicap, the human being must unhesitatingly be given priority without any attempt to establish that it somehow remains higher on the scale of moral assessment than the non-human.

Alluding to a well-known discussion of Bernard Williams,[14] they suggest that the attempt morally to justify giving preference to the human being by showing that it does emerge, after the matter has been carefully considered, higher up the common scale than the non-human, is to offer 'one thought too many'. The sole moral justification we need is 'that is a human being'. Human beings do not forfeit this claim on us however poorly they rank on any scale of admirable qualities, shareable with non-human animals, which we may choose to construct.

Lynch and Wells, therefore, see the attempt by defenders of ecologism to establish that human beings have the highest degree of intrinsic value and moral considerability as an implicit recognition of the truth of their claims. However, this approach, they suggest, does not work, for it will only be a contingent matter that any given human being does possess to a higher degree than any non-human the properties in virtue of which intrinsic value is attributed to a living creature. In a clash-of-basic-interests example, ecologism could thus be committed by the logic of its arguments to give preference to a non-human over, say, a mentally handicapped human.

Does this argument, as they believe, strike a killing blow against the approach outlined in the previous section?

First, we should notice that what they are implying in their use of the term 'human being' throughout their discussion is not intended to have any biological significance. That is, they claim that it is not because they are members of the same biological species that human beings justifiably give each other priority, so the position is not aptly referred to as 'speciesism'. The motivating thought is meant to reflect what they refer to as a 'fundamental modality of moral concern'.[15] In elucidation of this they provide the analogy of someone's justifying helping another because 'she is my sister'. This involves pointing to a special relationship between the two people which in itself is held to justify the one helping the other. No further feature needs to be alluded to in justification. Any such justification would once again be a thought too many.

What is the nature of this relationship, then? Does it provide a good analogy for the moral relation between human beings? One point which it is important to note is that the original discussion of such relationships given by Bernard Williams aimed to show, not that certain human relationships, such as those between spouses, are morally privileged, but precisely that some human relationships can have such importance for the lives of the individuals in them that they can lead to and justify behaviour, such as saving one's wife rather than another person, in the face of what moral reasoning would require.

That is, a man's unhesitatingly saving his wife is not motivated in such circumstances by a moral thought at all, but by something more fundamental in the life of the people involved. In this case it is, obviously, love which is at issue, as it usually is in the parent–child relation which provides another rich source of examples. This kind of relation is more fundamental than morality because it is essential to the meaningfulness of one's life, without a sense of which neither morality nor any other weighty consideration will be able to get a grip on one. Love in this sense is not just caring and concern, although it involves that, but a deep and direct connection with another in a way which creates a kind of fusion of the individuals concerned, as famously celebrated by Plato in the *Symposium*.

This is clearly a different kind of phenomenon from the one which Lynch and Wells describe. In their example, involving sisterly relations, they explicitly speak of the necessity for one to help the other as something which 'presents itself as a moral obligation'.[16] However, if it is a moral obligation, then it will be expressible in perfectly general terms, such as 'All sisters ought to be prepared to help each other in situations of dire need'. It is this kind of universalised statement of moral duty, used to justify help given, which Williams claims to be 'one thought too many' in the cases he considers. What a wife wants and expects to hear from her spouse in justification of his saving her rather than another, perhaps more estimable, woman is something of the form 'I love her and cannot live without her', not 'Husbands owe a general duty of care and concern to their wives'.

The example which they give of the spontaneous and unthinking saving of a human being under attack by a non-human animal seems to be akin to the 'wife' example, rather than to their 'sister' example. It is not presented as the outcome of any moral deliberation, but as a spontaneous manifestation of human solidarity in a situation of dire need. Plainly, there are good reasons why such spontaneous responses should be present in social animals such as human beings in situations of direct contact between them. Something like a deep-rooted biological basis for spontaneously helping a human being who may be a complete stranger is going to have to do the job done by the deep personal relationships cited in Williams' example. If there are such impulses, then this may explain the strength of the intuitions upon which Lynch and Wells are obviously relying in their argument. But if this is akin to the 'wife' case, then what it is referring to is something beyond morality, rather than an example of morality in action.

However, when this example is explained by Lynch and Wells, they do not see it as something which transcends morality, but put it squarely in the category of moral thought. They then seek to elucidate its basis by speaking variously of all human beings as having a special relation to all other human beings, *qua* human beings; of their having a fundamental attachment of care and concern to other human beings *qua* human beings; and of all human beings as forming part of a community of care and concern.

In so far as this takes the matter out of the 'beyond morality' category of examples, and puts it into the moral area under the heading of 'special relationships', this move immediately starts to open the Lynch and Wells position to the charge of being question-begging. It is certainly correct that, within moral thought, there is the need to establish special obligations between moral agents and between such agents and others, with respect to specific relationships. However, it is in principle a matter for debate as to what special relationships should be established and which obligations are involved in them.

For example, child-rearing requires that children be the special responsibility of some adults or other during their childhood. The obvious adults are the biological parents, which aligns biological forces with social needs. However, in some cultures it is the mother and mother's brother who do the rearing and have the responsibilities. A male parent thus will have special relations and obligations to nephews and nieces, rather than to his own offspring.

Thus, the morally charged language of 'communities' and 'special relationships' is open in principle to debate and the offering of alternative suggestions. This, of course, is precisely what ecologism is doing. It suggests that the 'community of caring' be extended beyond just the human realm. To reject this as in some way running counter to the very nature of morality is to beg the question at issue. The suggestion of granting a special privilege to the human community gains its plausibility from examples, such as saving the man from attack by a non-human predator, which draw their force from the fact that they are appealing to deep, morality-transcending impulses and feelings. Once detached from this context, the suggestion may be seen to express no more than the moral orthodoxy which is precisely what is under challenge.

That this is so may be gleaned from the fact that, outside of situations in which human beings are in direct contact, and needs can be directly registered, there is not a great deal of evidence that human beings do see themselves, outside the pages of moral theorising, as forming a community of care and concern. Our mutual destruction of each other – carried to extremes in modern forms of anonymised combat – and our daily ignoring of the plight of millions of children who die of hunger and easily preventable diseases suggest that there is no very strong current of moral feeling running in this direction. This, it will be recollected, is the claim made by Ruse in his elucidation of Darwinist ethics.

Of course, this may do no more than show that we are moral reprobates a great deal of the time, not that we should not accept the idea of a special community of care and concern which privileges its members over non-members. However, the key point remains, that once we detach the arguments of Lynch and Wells from examples which are properly to be construed as morally transcending, rather than examples of moral thought in action, then the arguments of the kind encountered in this and the previous chapter do require to be reckoned with, rather than to be dismissed as in some way missing the point of morality entirely.

What, then, should we say about the example which seems to pose the strongest challenge to the moral approach of ecologism? This is the idea that, when two creatures are in a situation of conflict of interest over a matter which concerns a basic need of both, one is a severely handicapped human being and the other is a non-human animal possessing greater intelligence and sensitivity than the human being, then the approach of ecologism commits it to giving preference to the non-human animal.

Various replies seem possible. One is that this could be encompassed by the Williams-style argument, that giving preference to the human being in those circumstances cannot be justified morally, but that morality is not everything. Another is to argue for the importance for human well-being that we do not kill or let die any innocent human being, for the psychological and moral effects of this would be deleterious. Another is that this case is no different in principle from the case where it is two human beings who are involved, the handicapped one and a healthy, normal one and where, if a choice has to be made, it is at least arguable that the healthy one be preferred. Dilemmas are difficult for all moral theories. A moral approach which seeks precisely to alter our picture of the moral universe will throw up new ones which at first sight will seem particularly hard to resolve. But ecologism is not more open to objection on this score than is any other moral theory.

In fact, the standpoint of ecologism is well placed to accept the argument that human beings should form a community of care and concern, as resting on the basis of what human beings need for their well-being, both as carers and cared-for. This is, however, compatible with the more general claim that moral obligations extend beyond that community. In this respect, the situation is no different from that involved in the familiar idea that members of families, neighbourhoods or nations form such communities, while their members nevertheless have wider obligations beyond these communities to human beings outside them.

In the light of the comments made in the paragraph before last, it is clear that ecologism inherits from its purely anthropocentric forebears the problem of establishing the case for a human community of care and concern. An attempt to show that it is not biologically impossible for human beings to achieve this was put forward in Chapter 3. But it is clearly not going to be easy. Moral theory has its part to play, and ecologism's moral theory is as well placed as any to foster this sense of community.

Justice, Contemporary Aliens and Future Generations

•

The moral theory of ecologism is maximally ambitious. It requires not only that we encompass the non-human in our moral thinking, but also the moral claims upon us of both contemporary human beings who are not members of our own societies and of future human beings. As is often noted, traditional moral theorising has largely neglected discussion of both of these areas of intra-human ethics. Many traditional theories are universalistic in orientation. That is, they are held to be applicable to some perfectly general class of being, as agents and patients, in virtue of some property possessed solely by the members of that class – personhood, rationality, capacity for experiencing pleasure and pain, and so on. But the full implications of this universality have only been fitfully explored. Overwhelmingly, moral theorising, even within universalistic moral theories, has focused on the moral relationships between contemporary human beings who inhabit the same society.

Undoubtedly this has been in large measure because for most of human history the persons whom we have been most able to affect by our actions have been our contemporary compatriots. Persons living in other societies, far away, have simply been beyond our reach and ken, and future people have been thought of as only minimally affectable by anything we can do, even as societies, within our own lifetimes. This has all changed, of course. With the spread of rapid communications and the increasing interconnectedness of the global economy and political system all human societies are in some form of direct contact and in a position to affect each other to varying degrees, directly or indirectly. Human technological impacts on the planet mean that we clearly now have the capacity to affect our descendants in dramatic and probably irreversible ways –

the effects of an all-out nuclear war being the example which comes most readily to mind.

Hence, for the first time in human history, moral issues which hitherto have been slumbering in the background of moral theory have come dramatically into the foreground as requiring urgent consideration. Of course, some traditional moral theories, those of Hobbes and Burke, for example, do not have the universalist import of the Kantian, utilitarian or other influential theories. They expressly restrict the meaning and scope of moral concepts such as rights and justice within social and temporal limits. For such approaches, the relations between societies is a matter of prudence rather than morality and the interests of future people is at most restricted to the need to maintain a 'contract' or heritage with or for future members of one's own society. Non-human nature, past, present or future, is, of course, not counted morally at all by such theories.

Universalist moral theories cannot be so restrictive without arbitrariness. Hence, within recent moral theorising we discover concerted attempts to apply moral concepts, such as rights and justice, to contemporary aliens and to future generations. Some universalist theories, such as utilitarianism, can, as we have noted, extend their moral reach to encompass at least some aspects of the non-human – roughly, to high-grade, sentient, individual non-humans. But by and large the focus has remained on human beings.

Within these universalist, humanist approaches, some reason has been found to restrict the extent of claims contemporary human beings within a given society may be required to acknowledge, even on universalist presuppositions. The most notable of these reasons is to be found in the concept of special obligations, to intimates and loved ones, which are held to restrict the wider claims which morality may properly place upon us as individuals.[17] Related reasons are that the claims of actual people have greater intrinsic weight than those of merely possible people; that there is an inherent limit, based on the nature of human psychology (compassion fatigue), on what sacrifices any individuals may be required to bear by morality; that the claims of morality in any case do not always properly trump other considerations (Williams' argument, encountered in the last section).

The concept of justice has come in for specific scrutiny in this area of moral theorising. The concept of international distributive justice, justifying the systematic transfer of resources across national boundaries from haves to have-nots, and the concept of intergenerational distributive justice, especially connected with the idea of sustainable development (handing on to future generations a package of resources no worse than the one we inherited ourselves), have emerged in this regard.[18] Here, too, there have been sceptical voices. Some theorists see the whole concept of distributive justice, at least as applicable to societies, as being open to objection and thus *ex hypothesi* object to new social applications of the concept.[19] Others hold that the concept only applies intelligibly to certain specific social situations – where the 'circumstances of justice' apply, or where reciprocity or rational co-operation are meaningful concepts. These exclude the application of justice, rather than humanitarianism, to the relations between contemporary non-compatriots and between present and future people, with whom it is impossible to enter into relations of co-operation or reciprocity. They also rule out as nonsensical the idea that non-human nature may have claims of justice upon human beings.

Let us consider specifically what ecologism has to say about the issue of justice between present and future human beings. In what follows I will use an analysis of theories of justice provided by Brian Barry in his discussion of this matter.[20] Barry first takes issue with a prevailing conception, associated particularly with Hume, but also lent credence by the use made of it in Rawls' prestigious theory, that justice only has applicability in a certain kind of situation, namely that in which the 'circumstances of justice' apply.

As Barry notes, these circumstances are summarised by Rawls as involving a situation of moderate scarcity of resources, moderate selfishness among human beings and relative equality (of the power to inflict harm) between them. Hume's analysis is intended to demonstrate that the concept of justice arises within a human society in which these circumstances apply as a result of the interest which members of the society have in acquiring and maintaining possessions. That is, justice concerns property – how it may be acquired and transmitted. The rules of justice which arise in these circumstances are conventions, with no justification beyond themselves, and form the basis for rational co-operation among members of the society. The rules are obeyed because, and in so far as, they safeguard the property of each on condition that each obeys the rules with respect to the possessions of others.

Barry demonstrates conclusively that the requirements of justice, specified in terms of 'giving to each person their due' obtain even outside the 'circumstances of justice' specified by Hume, so that the latter should be rejected as a prerequisite of the intelligibility or practicability of justice.[21] He usefully contrasts the Humean conception of justice as 'rational cooperation' with the Kantian conception of justice as 'universal hypothetical assent'.[22] Thus, he explains, the essence of Hume's analysis is that justice is what would be chosen by rationally self-interested people characterised by approximate equality in a situation in which they are able to gain the benefits of co-operation if they can devise a means satisfactory to them all of doing so. By contrast, the essence of Kant's theory is that the principles of justice are constituted by a hypothetical choice made by an individual under conditions that ensure that his choice has universal validity – that is, the principles would be selected by all rational choosers in that situation. The conditions in question are, of course, those which systematically preclude the chooser from biasing the principles to favour his or her own situation, an idea familiar from the Rawlsian device of requiring principles of justice to be chosen behind a thick 'veil of ignorance'.

Barry usefully points out that these rival conceptions converge when we consider how justice should apply to a 'self-contained society of contemporaries', precisely the situation to which traditional moral theory, as we noted above, largely restricts itself.[23] They begin to give divergent answers when they are applied to contemporary aliens, different generations of the same society and different generations world-wide. The possibility of rational co-operation between members of a given society and members of one or other of these groups is, of course, either non-existent (future generations) or very attenuated (with respect to many contemporary aliens). Thus, the Humean approach finds it impossible to admit that requirements of justice operate between such groups – benevolence and humanitarianism will have to take its place. This immediately brings in a significant weakening of moral requirements and, if the concern for members of such groups is founded on the fact of our caring about them, such requirements are rendered worryingly contingent – we may cease to care.

The Kantian theory has no such problems, since the principles of justice are not grounded on the possibility of rational co-operation and are not bound by considerations of time and place (as is true also of utilitarianism). It is possible in principle, therefore, to apply the Kantian approach across time and space. In its Rawlsian version, this amounts to a thickening of the veil of ignorance behind which principles of justice are to be chosen so that choosers are to be ignorant of the generation and of the particular society of which they are to be members when the veil is lifted.

Barry concludes his discussion with various observations. First, he says that the Kantian approach leaves it unclear why principles chosen in the manner stipulated should be regarded as principles of justice at all. The theory must be presupposing that there are 'constraints of right' which specify the range of morally arbitrary contingencies which have to be excluded in the formulation of principles of justice. He suggests that it is very unclear what such constraints might be, or how they are to be ascertained. In any case, he claims, the Kantian theory ends up by excluding too much as morally arbitrary. The details of human social life and interpersonal interaction – praise, blame, resentment, gratitude, compassion – are jettisoned in the Kantian approach, and with them go the necessity for the 'desert- and need-based' conceptions of justice which, he implies, we do require.[24]

He also suggests that the rational co-operation model, although it does not help us with issues of justice between generations and with respect to contemporary aliens with whom we cannot co-operate, does capture much of what we wish to encompass under the heading of justice.[25] It cannot be the whole story even here, though, for it cannot, within its own terms, establish the justice of the initial holdings of possessions upon the basis of which rational co-operation is possible. There is, however, an issue of justice here and we therefore need some other basis than that of rational co-operation upon which to determine it.

Barry suggests that the key conception is that of 'equality of opportunity'.[26] This appears to mean that the initial distribution of holdings must be such as to afford every individual equality of opportunity, although this concept is not further specified. With respect to justice and future generations, it produces the general requirement that 'the overall range of opportunities open to successor generations should not be narrowed'.[27] Barry gives an example drawn from environmental matters of what this might mean – 'If some openings are closed off by depletion or other irreversible damage to the environment, others should be created (if necessary, at the cost of some sacrifice) to make up'.[28]

Thus, the requirement of equality of opportunity is construed negatively – justice, he suggests, requires that we do not make our successors worse off. It does not require that we make them better off. Many who have held this kind of view have tried to deal with the obvious problem which this raises, of how we can be sure that our successors will hold the same view as we do about what opportunities are important ones and what counts as being better or worse off. Barry makes the observation that we do not need to know, or approve of, our successor's tastes in order to satisfy the requirements of intergenerational justice.[29] This suggests that the opportunities we are to preserve are specifiable in some way independently of specific tastes and preferences. Perhaps some concept similar to Rawls' conception of 'primary goods' is necessary here.

Barry's discussion establishes some important positions for ecologism. First, the rejection of the doctrine of the 'circumstances of justice' is very important. As Barry himself notes, the circumstance of approximate equality in the ability to inflict intentional harm between the parties to the convention of justice effectively rules out from that convention any beings incapable of such intentional harm directed towards the parties. This, he argues, would cover aboriginal peoples subject to the overwhelming power of colonial powers, and also rules out non-human nature from being a possible subject of justice.[30] It also, of course, appears effectively to exclude future people from being a subject of justice, since they cannot harm existing people,[31] although existing people can certainly harm them (even if we accept the case several have made, that the specific people who are brought into existence are in no position to complain of such harm, since without it different circumstances would obtain and a whole different set of people from them would exist instead).[32] Since ecologism would certainly wish to cover all of these groups with the cloak of justice, it is important that this block to doing so be removed.

Secondly, it is important that the rational co-operation interpretation of justice be restricted in scope. On Barry's interpretation its function is at most to determine, albeit only in part, the requirements of justice between contemporary compatriots, or perhaps between all contemporaries, compatriots or not, who are capable of rational co-operation. It cannot help determine the requirements of intergenerational justice, for between distantly separated generations co-operation, rational or otherwise, is impossible. Thus the rational co-operation model is insufficient for determining justice even between the only rational co-operators of whom we are aware. This is helpful for ecologism, which wishes to see the non-human encompassed by relations of justice, for the theory of justice which most obviously threatens to restrict justice purely to (human) persons is now seen to be inadequate even for persons.

It is also useful to have attention focused upon the issue of the justice of the initial distribution of holdings and the question of whether that distribution is able to sustain equality of opportunity or not. Given that the topic is justice, this equality of opportunity must be construed as what is due to the subjects of justice, not simply as what it would be nice or pleasant for them to have. Equality of opportunity also has the useful property for ecologism of being a concept which can readily be applied to all living beings.

It allows the thought that prima facie all morally considerable entities are to be accorded equality of opportunity to exist, flourish and develop in accordance with their natures. This is a prima facie case, of course. Some specimens and species may need to be deprived of such equality of opportunity so as to secure it for others, given that living beings interfere with the conditions of each other's existence and flourishing. But ecologism will certainly be happy to follow Barry's broad-brush approach to this requirement and to lay emphasis on the negative version of the requirement – we should leave future generations of humans (and non-humans) no worse off than we found them, bequeathing an environment in which their opportunities for existence and flourishing are no worse.

This leads on to a consideration of Barry's rejection of the Kantian approach to justice as involving some mysterious 'constraints of right' and as jettisoning too much of what is distinctive in human life. This is a curious position to take in the light of the observations he goes on to make, and which we have just been discussing, about the inadequacy of the rational co-operation model, even for human beings. For the attempt to find a more basic starting point for justice in the realm of 'initial holdings' is about as abstracted from the details of human life as it is possible to get.

The point of pondering the justice of initial holdings, of course, is not that we are seeking to establish any such holdings in the name of justice. It is far too late for that. The point of the conception is to find a theoretical point from which it is possible to criticise the pattern of actual holdings as having an unjust foundation, and therefore as justifying a rectification of injustice, perhaps by transfer of holdings, payment of compensation, or restitution of an initial condition. The principle of equality of opportunity when put to this use appears to be a paradigm case of a principle of justice whose justification rests on the thought that it is what would be chosen by a rational chooser in a situation in which all morally irrelevant considerations had been systematically excluded. Also, although Barry discusses the question of the justice of initial holdings in connection with justice and future generations it is plain that precisely this issue can be raised with respect to justice and contemporary aliens and present and future non-humans.

It may well be correct that the Kantian approach to justice is not adequate to give a proper account of justice between contemporary compatriots, and that some version of the rational co-operation theory does a better job of that. But the structure of Barry's own argument suggests that the Kantian approach may be foundational to the whole business of constructing an adequate theory of justice. Perhaps, then, Rawls is not as confused as Barry accuses him of being. It may be that he uses both the Humean and the Kantian approaches because he senses that they are complementary.

Of course, it is clearly true that we are in some sense in this abstract realm operating with notions of 'constraints of right', and it appears that such constraints always involve some postulate of equality. It is also clear that justifications come to an end sometime and that we have to treat some propositions as starting points in any argument. The Darwinian theory applied to the emergence of human moral thought explains why we treat propositions concerning the intrinsic value of human life in this manner. Whether or not this is the correct explanation, all moral theory has to begin with some intuitions about what is 'right', and it is no objection, therefore, to the Kantian approach that it expressly does so. It would be an objection if the starting points were morally arbitrary, but it is hard to see how a theory which aims to secure the equitable treatment of all human beings can be accused of that.

However, if we endorse something like Barry's account of the more fundamental point of justice as resting on a principle of equality of opportunity, and, as supporters of ecologism, we interpret this as applying to all morally considerable beings, we are left with some difficult issues to resolve. For example, with regard to future generations, Barry appears to suggest that we should not be regarded as acting unjustly with respect to them if we leave them with as wide a range of opportunities as we have, but thereby leave them with a narrower range of opportunities than we could have left them with if we had acted otherwise. For he tells us that justice does not require us to leave them better off than we are, but to leave them no worse off than we are.

Clearly this is, in a way, to respect an equality requirement, for they and we possess equal opportunities. But if the amount of opportunity for human beings has been steadily increasing through history, so that each generation has been leaving its successors better off, and we can do so too, it might be argued that we have a duty to act towards our successors as our predecessors acted towards us. This would not be an egalitarian principle, but might be construed as a requirement of fairness. This raises the question of whether, in effect, we are allowed to 'flatten' the curve of increasing opportunities across generations by taking from our successors resources which improve our opportunities but diminish theirs. They may still be left no worse off than we end up being when we do this, but less well off than they would have been had we not done it. It is possible that if we do not 'take' from future generations in this way our opportunities will not diminish, and may even increase, although not by as much as they will if we do 'take', so that it will not clearly be the case that we are acting unjustly to our contemporaries if we do not take. Are we allowed by Barry's principle to 'take' without flouting justice?

However we answer such conundrums, it is at least clear that we can meaningfully talk of the requirements of justice beyond the traditional arena of our contemporary compatriots, that theories of justice which apply to all the human beings to whom we wish them to apply – contemporary aliens and future generations – can only be justified by very abstract moral postulates of equal moral considerability, and that such postulates are then applicable to the non-human too. This is another application of the idea that if we develop a moral theory that applies to all human beings, we find that it applies to more than human beings.

The issue of justice with respect to future generations has recently been given a distinctive interpretation in terms of the concept of sustainable development. This is a concept to which it will be appropriate to return when, in Part Four, we turn to consider issues of political economy.

Notes

1. Will Kymlicka, *Contemporary Political Philosophy* (1990), p. 5.
2. Nicholas Low and Brendan Gleeson, *Justice, Society and Nature* (1998), p. 2.

3. Low and Gleeson 1998, p. 2
4. Freya Mathews, *The Ecological Self* (1991).
5. Mathews 1991, p. 123.
6. It will be seen that this argument may also be used to elucidate different degrees of wonderfulness. Degree of wonderfulness is determined by, *inter alia*, degree of complexity.
7. This example and argument in which it is employed are on pp. 123–6 of Mathews 1991.
8. Mathews 1991, p. 127.
9. A claim also made by H. Rolston in *Environmental Ethics* (1988), pp. 223–4.
10. See Dale Jamieson, 'Animal liberation is an environmental ethic' (1998). He operates with a schema showing similarities to the one employed here. His comprises two value-distinctions – intrinsic/instrumental and primary/derivative – and one value dimension: greatness or urgency. I think it is very hard to see any difference between the first pair of distinctions. Accordingly I operate here with the distinction between intrinsic and instrumental value and two value dimensions: degree of intrinsic value and urgency of value-claim.
11. Douglas Booth, 'Preserving old-growth forest ecosystems' (1997).
12. Booth, p. 43.
13. Richard Routley and Val Routley, 'Against the inevitability of human chauvinism' (1979).
14. Bernard Williams, *Moral Luck* (1981), p. 18.
15. Tony Lynch and David Wells, 'Non-anthropocentrism? A killing objection' (1998), p. 157.
16. Lynch and Wells 1998, p. 257.
17. See James Fishkin, *The Limits of Obligation* (1982).
18. On international distributive justice see, for example, Charles Beitz, *Political Theory and International Relations* (1979). Sustainable development will be discussed in Chapter 10.
19. Most notably F. Hayek, *The Constitution of Liberty* (1960) and R. Nozick, *Anarchy, State and Utopia* (1974).
20. Brian Barry, 'Circumstances of justice and future generations' (1996).
21. Barry 1996, pp. 207–28. In sum, Barry shows that issues of justice – giving to each what is their due – intelligibly arise in situations other than those which Hume specifies as necessary for such issues to arise, and that when they do so they rest on a basis other than that of self-interested utility, which is the only motive countenanced by Hume. In any case, individual self-interest as the motive supposed to underpin justice is defective, as the example of the 'self-interested knave' (p. 217) shows.
22. Brian Barry 1996, p. 234.
23. Brian Barry 1996, p. 237.

24. Brian Barry 1996, p. 241.
25. Brian Barry 1996, p. 243.
26. Brian Barry 1996, p. 243.
27. Brian Barry 1996, p. 243.
28. Brian Barry 1996, p. 243.
29. Brian Barry 1996, p. 244.
30. Brian Barry 1996, pp. 220–1.
31. However, John O'Neill, in *Ecology, Policy and Politics* (1993), pp. 28–36, argues that future generations may harm present ones by failing to bring, or failing to be capable of bringing, to fruition the projects which give meaning to the lives of the present generation. This establishes a duty on each generation to secure a shared community between generations. This is a powerful argument, although it does not help with the cases of contemporary aliens or the non-human, with which relations of community, in any meaningful sense, are often absent. Barry's argument against basing justice on capacity to harm intentionally is still needed for these cases.
32. See Derek Parfit, *Reasons and Persons* (1984), pp. 357–61.

Part Three

POLITICS

6

Ecologism's Political Philosophy: Human Nature, the Human Predicament and Political Morality

LET US ASSUME THE correctness of the claim that the current biodiversity of the planet is under severe threat as a direct result of human activity and that the moral case has been established for saying that we ought to take action to halt and even reverse this process, in order to take proper account of the moral considerability of the non-human.

Before we can proceed to a consideration of practical measures to achieve these aims there is another level of normative theory to explore – the political. We need a worked-out political philosophy which gives us the normative arguments for organising our political decision-making in such a way as to integrate into it the moral requirements we have identified.

Luke Martell has explicitly argued that, while ecologism has reoriented political philosophy by introducing the new moral issue of human/nature relations and the moral considerability of the non-human, it needs to draw on other philosophical traditions in order to answer key issues about equality, liberty and justice within human societies.[1] Ecologism is held not to have anything distinctive to say on these issues on the basis of its distinctive moral premises. In view of this claim we must try to decide what form ecologism's political theory should take. Must it be a hybrid, or can a distinct theory be formulated which answers questions about human relations and socio-political arrangements from a viewpoint significantly different from those of traditional political philosophies?

Let us begin by considering what ingredients a political

philosophy needs to contain. Traditionally, the dominant political theories of the West have comprised the following:

(1) A theory of human nature and the human predicament, showing why some form of political organisation is necessary in order to do justice to that nature and to cope adequately or well with that predicament.
(2) A theory of political morality, setting out the basic moral concepts to be deployed in the elaboration and implementation of political arrangements and aiming to demonstrate which moral concepts are basic and why; how they are to be deployed in moral reasoning; what political (and socio-economic) arrangements they mandate or permit; which possible forms of human political arrangement are to be regarded as the best, morally speaking.
(3) At the meta-level, theories concerning how moral thought is to be understood, moral judgement and argument to be construed; about how one may arrive at the most defensible views of human nature and the human predicament used in (1); about how the normative discourse of morals and politics relates to other normative discourses (such as religious ones) and to non-normative theories, such as those of natural and social science.
(4) A system of political economy, which shows what modes of economic activity are required by, or are at least compatible with, the underlying moral system and its concomitant political order – specific to this area are discussions of praxis, human individuality and distributive or social justice.
(5) Attempts to employ the findings of political science, economics and sociology in order to refine the concepts of what is feasible in the creation of human political systems, where weaknesses and dangers may lie and what steps may practically be taken to obviate these.
(6) Given political philosophy's orientation towards practice, attempts to operationalise its key concepts, to explain how to get from the existing situation to the preferred one by morally acceptable steps, and to explain how the preferred situation is to be maintained by morally acceptable means in the face of actually or potentially destructive forces. Some theories aim to explain what to do if they should turn out to be incorrect – what mechanisms of alteration and reversal they countenance – and how far ideal and reality may be expected to coincide.

These issues will be addressed in this and subsequent chapters. In each case we must try to pinpoint those views which may be regarded as distinctive of ecologism.

Ecologism's Theory of Human Nature and the Human Predicament

As we have already noted in the discussion in Chapter 3, ecologism takes a distinctive view of this set of issues. Let us recall the general outline of the position. Ecologism emphasises as fundamental the idea that we are an animal species. This immediately implies that, as is the case with other species of animals, our history and characteristics are now best approached via a grasp of the theories of the modern biological sciences, especially the neo-Darwinian theory of evolution and the science of ecology.

Thus ecologism is naturalistic and science-oriented. It begins with a view of human beings as natural beings and views them as subject to the same imperatives of survival and competition in a world of limited resources as apply to all species. It views them as having co-evolved with their fellow life-forms on this possibly unique planet, as sharing genetic material with them, and as being related to them as elements on a single continuum, rather than as being separated from them by some unbridgeable gulf. In these respects ecologism commits itself to a controversial view and one which, depending so heavily as it does on scientific theorising, may yet turn out to be incorrect. Thus, an acceptance of this starting point of ecologism is something of a gamble.

Other claims encountered in Chapter 3 to which this naturalistic/science-oriented starting point leads are that our self-understanding is best approached via an integration of the insights gained from evolutionary theory, genetics and ecology with the findings of the social sciences. The project of sociobiology is bound up with this. Even if sociobiology is ultimately rejected, the basic claim of ecologism will remain in some form, that the clear and rigid distinction between natural and social science is no longer tenable.[2] A further specific claim of importance to ecologism is that of Wilson, to the effect that there is the phenomenon of 'biophilia', or human love of the biosphere, whose content is to be understood best via a grasp of our evolutionary history.

In addition to these points about our 'inner' nature, ecologism's naturalistic/science-oriented premise leads to distinctive claims about our 'external' situation. Specifically, the claim is made that our fundamental predicament is best grasped via the concepts of 'ecosystem' and 'biosphere'. These depict our human life-support system as a single interconnected whole in which our existence, well-being and the potential improvement of our condition as a species are closely bound up with the well-being and existence of millions of other species.

A further important claim about the human predicament drawn from this approach is that there are limits to our capacities for manipulation of our natural environment, limits which derive from what is needed to maintain the overall health and functioning of the biosphere. These limits are ones of which we have only recently become aware, just at the point, in fact, where we are about to run up against them.

We noted under (1) above that the traditional point of a theory of human nature and the human condition is to provide a basis for the justification of the political organisation of human beings. Let us consider whether ecologism's starting point has a similar purpose. On one view it does not. On this view the point of the ecologists' claims about human nature and the human predicament is a limiting or cautionary point, not a foundational one. That is, it is supposed to show that, whatever the point of our organising ourselves politically we must never lose sight of the way in which our situation as animals in a biosphere conditions and limits our political practices.

On another view, however, the ecologist's starting point can perfectly well be used for foundational purposes, to provide a distinctive rationale for political organisation. It can do so by showing, in the manner of many traditional arguments in political philosophy, why the absence of human political structures, or the presence of natural anarchy, will prevent human beings from dealing with the serious, and life-threatening, vicissitudes of life as an animal species inhabiting a biosphere. In this regard the 'tragedy of the commons' plays the same role in ecologism as the version of the prisoners' dilemma we find in Hobbes' *Leviathan* plays in his theory.[3] Thus, it can be argued that with no political organisation around to regulate and control human interaction with the biosphere in the course of meeting life-needs, human beings will deplete the life-support systems, sooner or later.

Of course, it is a matter of controversy as to what specific political conclusions may properly be drawn from this kind of argument. Supporters of private property regimes and free marketeers will resist any attempt to use such a conception to defend strongly statist conceptions of human political organisation. But even such theorists are usually 'minimal state' advocates, not anarchists. They accept that the state is needed to defend the private property and free market institutions.[4] Thus, if these in turn are given a defence in terms of being the best way to defend the biosphere against the tragedy of the commons, then they have in effect accepted the logic of ecologism's case.

There is, of course, a significant difference nevertheless between ecologism's justification of political organisation and that of traditional theories. This is that ecologism's justification conceives of the harms and benefits relevant to the justificatory argument to be harms and benefits to the ecosystem of the planet, which, of course, includes human beings but does not concern them alone. By contrast, the traditional theories look to politics for the securing of benefits/avoidance of harms in ways which do concern human beings alone.

Thus, as we have been discovering in Part Two, ecologism requires that the justification of politics takes account, for non-instrumental reasons, of the interests and well-being of the non-human. That is, in justifying a form of politics, ecologists require the political system which they wish to defend to be systematically and permanently committed to taking into account the well-being and conditions of flourishing of non-human creatures.

Thus, the key point to emphasise here is that ecologism's view of the basic dilemma of politics for human beings is not the traditional 'How can we human beings secure the benefits of our co-operation while minimising the drawbacks?' but 'How can we human beings secure for ourselves and for the rest of the biosphere the benefits of human co-operation while minimising the drawbacks for ourselves and the rest of the biosphere?'

Another way of formulating the same point is to think of the traditional 'state of nature/social contract' nexus as involving human beings as acting not just for themselves but as guardians or proxies for beings which it is impossible to conceive of as entering into a contract or dialogue – their fellow-creatures on this planet. Their motivation for doing so is held in part to derive from some sense of interconnection, solidarity or mutuality with those creatures – biophilia – and in part from a recognition of their moral considerability. To put the point in Lockean terms, it is seen to be a natural duty, part of the natural law, to preserve the biosphere, not just 'mankind'. The implications of this 'guardianship' standpoint for the institutions of human political organisation will shortly be developed.

It is an important point of the science-based analysis of human nature and the human condition to demonstrate why such a view is not wholly outlandish, but is rather to be regarded as entirely reasonable for an intelligent animal with our evolutionary background.

Ecologism's Political Morality: The Contextual Self and Moral Guardianship

For ecologism, then, the moral considerability of the non-human plainly has to be brought in right from the beginning and has an impact on the key question of what other political/moral values are to be aimed at by the political system.

To explore this claim further, we may begin by noting that Thomas Hobbes saw the main point of government as being the provision of security for individuals. John Locke, in his seminal argument for what has become the liberal view of politics, argued that the great aim of government is the protection of individual life, liberty and estates. For the tradition of socialism the point of government is to secure the conditions for the all-round development of the individual in a context of social solidarity and mutuality.

We can here see a movement of thought in which the individualistic perspective explicit in Hobbes' corpuscularian analysis of human beings is in due course challenged by a view of human beings as essentially contextual beings. On the latter view, the conditions for human well-being are to be found in a proper appreciation of human individuals' interconnection with other phenomena – specifically, in the socialist (and conservative) traditions, with others of their kind.

Ecologism introduces a different conception of human beings as contextual beings, whose conditions of flourishing depend on their relationship with a wider context. The wider context according to ecologism is, of course, provided by the biosphere and the myriad life-forms which make it up. This context was not ignored by previous philosophical traditions, but it was understood in a very particular way, namely as the source of the relatively scarce resources within which human beings, as fragile physical creatures, have to seek their livelihood.

This understanding, although correct as far as it goes, is, as far as ecologism is concerned, woefully inadequate. Human beings certainly are physically embodied and need the right kind of material setting within which to find the conditions for their continued physical health and existence. But they are also essentially beings with a moral sense, creators of cultures and possessed of a sense of love for the natural world. In these three areas an important dimension of analysis has been neglected.

It is pertinent in the context of the development of ecologism's political philosophy to say a bit more about the third at this point. The phenomenon of love of the living world reveals that human beings have a direct emotional and self-understanding relationship with their natural context. This aspect of human life, the crucial fact that human beings have a capacity for love, is, of course, not neglected in other political philosophies, even though it is curiously neglected in many expositions of them.

For the liberal it is the love of human individuals for each other within close personal relationships that occupies centre stage. For conservatism, love of country, understood as encompassing land, culture and history, is just as essential, and is viewed as wrongfully neglected by other philosophies. For socialism, love of humanity, reaching beyond the limits of particular sub-groupings, and expressed in human solidarity and fraternity, is essential to human life's flourishing. For religious philosophies, all these loves are to be found as aspects of the love of God. Ecologism, as already emphasised, points to the love which is directed to the living world around us as being of at least as central concern as all the others just mentioned.

Of course, this is a normative claim as much as a factual one. It involves the view not just that as a matter of fact such love is central to the lives of most human beings, but that a fully rounded human life should contain a place for such love within its emotional structures. To the extent that a given human being has no such love, then to that extent his or her life is impoverished. It may, of course, be that other forms of love are strongly present, and in sufficient degree to make the life lived a flourishing one from the person's subjective viewpoint. But still, on ecologism's view of these matters, there will be an important dimension missing from the person's life.

It is, of course, a matter of controversy between the upholders of the various political philosophies whether all of these forms of love are feasible, meaningful and essential to human flourishing. Socialists are sceptical about the role of patriotic love; liberals view 'love of humanity' with a certain amount of suspicion; religious philosophers view the love of humanity central to secular philosophies as dangerous when divorced from the love of God. All these forms of love can be viewed as snares to trap people into unlovely forms of life. They may in some cases be plausibly presented as in reality disguises for certain kinds of hatred. The love of one's natural context and other life-forms, for example, is often viewed in just this light, as a disguised form of misanthropy.[5]

These criticisms of the different forms of human love are often telling. But they reveal, not that such forms of love should be jettisoned or downplayed, but that they are powerful elements in human life which can be turned in destructive directions unless their precise nature and importance are clearly seen.

To return to the main point of this discussion, then, ecologism seeks to enrich our conception of ourselves as contextual beings. As we noted in the survey of ecologism's moral theory in Chapter 4, environmental ethicists have often spoken of the need to work out an 'extended' sense of ourselves as having direct and ethically important connections with the non-human natural environment. A recent case in point is the discussion by Low and Gleeson of the necessary underpinnings for a sense of justice as extending to our treatment of the non-human realm. We must jettison the 'bounded self', as they put it, of traditional liberalism-cum-utilitarianism, in which we view ourselves as individual preference-satisfiers, and come to see ourselves as having a variety of direct and essential interconnections with human and non-human others, such that the relations between ourselves and their selves can be clearly brought under the rubric of justice.[6]

What follows from these points is, then, that for ecologism the point of government is to secure the conditions for human flourishing, but in a way that does justice to our context-bound nature as beings with an essential moral, emotional and cultural connection with the non-human world, from which we also have to secure the conditions of our physical existence. The form of government which will be satisfactory for such creatures will be different in various ways from the forms countenanced by traditional political philosophies.

First, it will be circumscribed in its activities not simply by the requirement to respect basic human rights, but also by the requirement to recognise the moral considerability of non-humans. This will introduce a whole new set of side-constraints upon what human beings may permit their governments to do and will establish a whole new set of goals. These are the ones connected with what is required to maintain biodiversity and the conditions for flourishing of a vast range of other creatures. The issue here is essentially that which in the last chapter, following Low and Gleeson, was dubbed the issue of 'ecological justice'. It concerns the principles for determining the allocation of environments between human beings and non-human beings.

This has direct implications for what ecologism has to say about the accountability of the governors to the governed, which underpins traditional discussions of democracy. For a means has to be found for ensuring that human governors attend to the interests of the non-human objects of moral concern. There is obviously no way for such non-human moral patients to be participants in political processes or exercise democratic power to exact the requisite concern on the part of human governments. What is needed, therefore, is some institutional arrangement within human government for holding it to account on such matters.

Low and Gleeson have recognised the force of this argument. Given their concern with the practicalities of securing ecological justice in the contemporary world they have posited the need for a pair of democratically accountable global institutions with, in part, the responsibility of enforcing the requirements of ecological justice between all the earth's inhabitants, human and otherwise. They suggest the creation of a World Environment Council and an International Court of the Environment, to be established under the aegis of the United Nations.[7] The former would provide a forum for political debate and public scrutiny of global environmental issues, the latter would adjudicate specific environmental disputes which have a clear international aspect. Within these institutions ecological justice would be safeguarded by human advocates acting on behalf of the interests of the non-human.[8]

Low and Gleeson suggest that the role of the Council will not be to make decisions on matters of environmental conflict, but rather to set out the principles for making such decisions and to create institutional means for making them in specific instances. What they have in mind with respect to the latter is 'a series of accountable authorities bringing together the interests concerned, much like the various multilateral committees set up under the aegis of the UN for the creation of regulatory regimes'.[9]

However, it is clear that the logic of the argument would also justify the setting-up, within national political dispensations, of constitutionally entrenched environmental Councils and Courts, performing a similar function, especially the enforcement of ecological justice, at the national level, with respect to environmental conflicts which do not go beyond national boundaries.[10] By parity of reasoning this institutionalisation process may even go down as far as the local level. The right of such bodies to exist and their real ability to deliver ecological justice both require that they be constitutionally entrenched. This will not make their abolition impossible, of course, but it will make it very difficult. This is vital, given the powerful political pressures to which they will undoubtedly be subject by the activities of wealthy and influential economic actors.

It might be argued at this point that if enough voters can become convinced of ecologism's case, then they will require that politicians running for office do attend to such matters, so that no institutional arrangements in addition to normal democratic voting will be needed to enforce the moral claims of the non-human. If there are not enough people convinced of ecologism's case, then no institutional arrangement for enforcing the moral consideration of the non-human will last in any case.

In reply it may be conceded that it is certainly plausible to claim that ecologism's preferred political structure will only exist and work if a majority of people accept the moral, scientific and cultural arguments of ecologism. How to cope with those sceptical of, or hostile to, that case within the form of politics preferred by ecologism is a crucial issue to which we will have to return. It is the issue of how an ideology is to cope with only partial compliance with its prescriptions, and encompasses the issues of socialisation, education, toleration and totalitarianism within ecologism's political philosophy.

However, even granted the conversion of majorities of humans within specific societies to the ecologism's viewpoint it is clear that just leaving the defence of the moral claims of the non-human to the normal processes of human democracy will be inadequate. There will always be a temptation, even in a world dominated by ecologism, to ignore, overlook or downplay the claims of the non-human in the name of some human concern which is presented as pressing. After all, this happens already when human beings in one society are considering the moral claims of their fellow human beings in societies other than their own. Hence the need for such constitutionally entrenched guardianship institutions charged with the duty of reviewing the political, economic and social activities of the society to ensure that the moral claims of the non-human are being met. This will require a power of review and a veto power with respect to legislation and policy proposals.

For this to work, of course, ecologism needs to have a clear set of principles for determining what are the moral claims which can intelligibly be attributed to the non-human, which requires in turn a careful analysis of the moral standing of different species and perhaps of habitats and ecosystems. Principles such as those developed in the previous chapter will be necessary.[11] The successful functioning of such bodies will also require that they possess the wherewithal to gather the environmental impact information which will be needed to gauge accurately just what the effects of proposed legislation and policy initiatives will be.

Something of this kind is already working in the USA, where the Endangered Species Act, interacting with the power of judicial review vested in the court system right up to the level of the US Supreme Court, is allowing courts to prevent some economic activities on the basis that they threaten the continued existence of endangered species.[12] But, of course, the law in question is not constitutionally entrenched.

However, to refer for a moment to issue area (5) referred to at the start of this chapter, concerning ecologism's account of how its ideas may be achieved in practice, this phenomenon suggests one way in which the ecologists' preferred political structures may be brought about – not by wholesale and immediate revolution, but by the piecemeal emendation and transformation of existing socio-economic and political structures and practices. Several advocates of an ecologically sensitive political order, such as Low and Gleeson and de Geus,[13] have mooted the possibility of revolution-by-increment in this kind of way.

Such a process, of course, has the great advantage that it avoids the unpredictability of violent or rapid revolutions, which are too unstructured to guarantee the attainment of the revolutionaries' aims. It has the drawback that it may be too slow to be effective in a situation of rapid change and it may allow the enemies of the revolution ample time and opportunity to organise effective opposition to the transformation. Arguably, however, short of an environmental catastrophe, ecologism has no alternative but to pursue this incremental revolutionary route.

Ecologism's Political Morality: Ecological Justice and Environmental Justice

•

The creation of human political institutions to embody guardianship of the moral standing of the non-human will undoubtedly call upon resources of human selflessness which have not been conspicuously present even when humans have been considering only the moral claims of their own species. But there is nothing incoherent in the idea of such guardianship. Nevertheless, there is an obvious difficulty for the idea of such a guardianship which stems from the moral claims of human beings to an environment within which they may flourish. This set of considerations has usefully been labelled 'environmental justice' by Low and Gleeson.[14]

A consistent moral position for ecologism clearly requires that the environmental conditions to sustain human flourishing be accorded moral importance. One of the glaring facts of our contemporary situation is that there is great injustice in the distribution of environmentally beneficial environments among members of the human species. The unjust phenomena particularly criticised by Low and Gleeson in their comprehensive discussion of environmental injustice comprise the distribution of LULUs (locally undesired land uses) within societies on the basis of class and ethnic criteria which are plainly oppressive; the transportation and dumping of toxic substances from societies where they are not desired to societies the rulers of which, for reasons of national and personal economic self-interest, are willing to see their societies used as dumping-grounds; and the investment by Multi-National Corporations (MNCs) in production units in societies which, again for economic self-interest, are prepared to see their factory and agricultural workers subject to environmentally less favourable working conditions than those required by law in the MNCs' own societies.

Some of these injustices can clearly be overcome without any injustice being done to the non-human. That is, in many ways ecological and environmental justice are clearly compatible with each other. For example, if landfill sites are necessary, and they have to be put somewhere, then the only consideration of justice may be how the burden of being physically close to them, assuming this is an environmental 'bad', is to be apportioned among human beings. The creation of such sites, especially if they can be kept to a minimum, may pose no threat to the well-being of non-human individuals, species or ecosystems.

However, some poverty populations may take the view that for them the requirements of ecological justice must take second place to the requirements of securing economic advancement or at least greater economic security than they can otherwise obtain. If acting as a dumping ground for overseas toxic waste is the only feasible way for a society to earn an income from abroad, then its inhabitants may feel they have no choice but to sacrifice environmental quality, including their own in the short term, to gain the cash they need for economic advancement. If this results in the destruction and poisoning of habitat necessary for the well-being of other species, then this may be regarded as the price which needs to be paid for such human economic advancement. The latter in turn may be viewed as the necessary step towards the creation of an environment within which the human beings concerned may flourish, and thus to the securing of environmental justice in the long run.

Many such arguments will be specious of course, and rest upon the self-interest of local elites who gain the cash but do not suffer the environmental bads. In many cases a more just international economic system, transfers of money and/or technical know-how from rich to poor populations, the creation of a more internally just society and so on will be ways of achieving environmental justice and economic advancement without sacrificing ecological justice. But it appears to be at least a theoretical possibility that the destruction of habitats and the extermination of non-human species may be necessary if a human population is to achieve economic advancement or security. As Low and Gleeson note:

> In truth the harshest contradiction is perhaps between the demands of environmental justice – the need to spread good environmental conditions, and those of ecological justice – the protection of the planet for the flourishing of all life forms and ecological systems.[15]

They do not explore the implications of this thought. Closer inspection of what is at issue in such cases, however, suggests that, properly conceived, there is no conflict of justice. To see this, we should recollect that ecologism, in situations of moral trade-off in which basic needs of human beings and non-humans clash, gives human interests priority, but insists that such situations must be carefully considered to ensure that the clash is genuine and insurmountable. The interests of the non-human are not negligible, and the onus of proof is put on those who would perpetrate what is prima facie an ecological injustice in order to secure environmental justice between human beings to establish their case beyond reasonable doubt. This means, of course, that all alternatives to the destruction of the habitat crucial to non-humans must be considered and evaluated. Undoubtedly this task will be a major responsibility of the 'guardianship' institutions mooted above.

However, if this is done then ecological justice will also be done. This is because the point of the prioritisation rules mooted in the last chapter to adjudicate clashes of interest between different life-forms is that when they are followed the outcome is in line with the requirements of justice, not in contravention of them. In other words, a genuine case in which the basic needs of different species cannot equally be met is one in which justice is done when the interests of the more morally considerable parties are given precedence. But such are the dire consequences for the less morally considerable parties in such cases that the establishment of the incompatibility is subject to very rigorous requirements of proof.

This means, in effect, that the requirements of ecological justice circumscribe the requirements of environmental justice. In looking for ways to achieve a just distribution of environmental goods and bads between human beings we must always seek those ways in which the habitat needs of non-humans are not threatened. Only when human basic needs are in unavoidable conflict with non-human basic needs may we sacrifice the latter for the former, and we must do our utmost to avoid such clashes.

Having established the case for political arrangements designed to do justice to the non-human, as well as to ensure that that subsection of human distributive justice known as environmental justice is also performed, we now need to develop more fully the ideas concerning political arrangements to which we have so far only alluded in this chapter. Specifically, we need to ascertain whether ecologism has specific things to say about the state, democracy, world government and the more theoretical meta-level issue underpinning them all, namely what its view has to be about how political philosophising is to be conducted. These issues will be addressed in the next chapter.

Notes

1. Luke Martell, *Ecology and Society: An Introduction* (1994), p. 139.
2. See Martell 1994, Peter Dickens, *Society and Nature* (1992) and David Goldblatt, *Social Theory and the Environment* (1996).
3. As Dryzek points out, the logic of Hobbes' 'war of each against all' in the state of nature is identical with that of the 'tragedy of the commons'. See John Dryzek, *The Politics of the Earth* (1997), p. 25 n.

   John O'Neill (*Ecology, Policy and Politics* (1993), pp. 38–9) has succinctly summarised the original version of the 'Tragedy' (put forward by Garrett Hardin in 'The tragedy of the commons', in G. Hardin and J. Baden (eds), *Managing the Commons*, San Francisco: Wm Freeman, 1977) as follows:

   > A pasture is open to a number of herdsmen, each one of whom acts in isolation from the others and attempts to maximise his own utility. Each considers the utility to himself of adding another animal to his herd. For any individual, the positive benefit of adding another animal to his herd will be greater than the loss from overgrazing, since the benefit accrues entirely to the individual, while the loss is shared among all the herders. Hence, a set of rational, self-maximising herdsmen will increase their herds even though collectively it is to the detriment of all.

   It is pertinent to remark, however, that, as O'Neill notes (p. 38), the designation 'tragedy of the commons' is a misnomer. The 'commons' in question are really open-access regimes to which no-one can control access, not resources which could be held in common and access to which could be controlled by, for example, a village commune.
4. Dryzek 1997, chapters 3 and 6, provides searching analyses of these kinds of response.
5. A misanthropy that has occasionally surfaced explicitly in the history of modern environmentalism. An account may be found in Dryzek 1997, pp. 156–7.
6. Nicholas Low and Brendan Gleeson, *Justice, Society and Nature* (1998), pp. 135–7. I have some doubts about certain interpretations of the notion of an extended sense of self – see my 'Ecocentrism and persons' (1996). A grasp of interconnections with, and a feeling of love for, non-human nature are the most defensible part of the concept. A sense of identity with non-human nature is much more dubious.
7. Low and Gleeson 1998, p. 190.
8. Low and Gleeson 1998, p. 191.
9. Low and Gleeson 1998, p. 191.
10. We can find a similar position, albeit without the mention of a specific guardianship institution, in John Barry 'Sustainability, judgement and citizenship' (1996), p. 122:

> Apart from embodying the present generation's obligations to the future, a constitution could be considered as expressing a society's considered and deliberate attitude to non-humans. That is, insofar as we can consider both non-humans and future human descendants as 'moral subjects'... a constitution can provide some legal protection for these vulnerables. There is nothing startling about this since such legal incapacities are common features of liberal democratic politics. We can think of this as involving constitutional provision for the representation of the interests of non-humans as well as future citizens.

Another author in the same volume who suggests the need for an ecologically informed democracy to have some legal/constitutional means to protect the non-human is Peter Christoff in 'Ecological citizenship and ecologically guided democracy' (1996), pp. 165–6.

11. On pp. 156–7, Low and Gleeson 1998 offer their own version of such principles.

    First Principle: Every natural entity is entitled to enjoy the fulness of its own form of life; non-human nature is entitled to moral consideration. With an extended conception of the self an absolute barrier between human and non-human is untenable.

    Second Principle: All life-forms are mutually dependent and dependent on non-life forms. This principle must be considered when any conflict among species occurs. Exactly what implications this has for judgements in specific instances of conflict between the rights and needs of different life-forms is as yet unclear.

    To these are added three 'rules of thumb' to govern their application:

    (1) Life has moral precedence over non-life.
    (2) Individualised life-forms have precedence over life-forms which exist only as communities.
    (3) Individualised life-forms with human consciousness have moral precedence over other life-forms.

    The latter rules obviously aim to set up bases for moral trade-offs of the kind suggested in the last chapter, although some issues, such as the relative importance of species and individual specimens, are not touched on in them.

    As noted earlier in this chapter, the first two principles really involve red herrings. Justice requires to be done to other life-forms even if we do not develop an extended sense of self and even if we are dealing with life-forms, on another planet, say, with which we have no interconnection (other than the bare epistemological one, that we know of their existence). Justice needs to be done to such beings because of what they are, not because of what we are, although, given what we are, it is we who need to do the justice.

12. See Bryan Norton, *Why Preserve Natural Variety?* (1987), pp. 3–6, for a presentation of this phenomenon.

13. See Marius de Geus, 'The ecological restructuring of the state' (1996), p. 188.
14. Low and Gleeson 1998, p. 2.
15. Low and Gleeson 1998, p. 183.

7

Ecologism's Political Philosophy: Political Morality and Meta-issues

FOUR ISSUES ARE IMMEDIATELY raised by what was said in the last chapter:

(1) Is there a distinctive case in the ideology of ecologism for democracy – may not an authoritarian system more efficiently achieve ecologism's aims? If there is such a case, does any particular form of democracy receive the support of ecologism?
(2) Does not the polity which ecologists are contemplating necessarily have to be conceived as global or international, given that the protection of the conditions for flourishing of life-forms is inherently a global issue? Must not ecological government be global government? If so, how can it be achieved and how can it be made to function satisfactorily?
(3) Does the traditional structure of government centring on states, conceived of as territorially limited entities claiming the sole legitimate exercise of coercive power within those territories, need to be replaced, from the standpoint of ecologism, with a looser conception of government, perhaps based on the developing network of global environmental non-governmental organisations? If so, how are problems of co-ordination across the globe to be dealt with?
(4) What is ecologism's meta-ethical position? Does it have to commit itself to any specific view concerning how to understand what is going on when persons debate ethical matters?

The last issue was touched on earlier in Chapter 3, when the naturalist approach to ethical thought was introduced in connection with biological approaches to understanding humanity. It is an issue which is particularly pressing for ecologism if it is committed to a global political dispensation, for then it will be necessary to address the claims, most recently associated with Postmodernism, of cultural relativism with respect to evaluative, epistemological and metaphysical matters.[1]

This fourth issue is particularly salient in connection with important aspects of the first two issues noted above, concerning problems of democratic decision-making in multi-cultural contexts and problems of global government and governance. Hence it will be discussed as part of the general issue of how trans-cultural agreement may be achieved on moral matters.

Ecologism and Democracy

We need first to recall the direct connection which ecologism posits between concern for the non-human natural world and concern for human well-being, which involves the development of an enriched concept of the latter. In considering the most desirable form of government, therefore, ecologism has to address the question of what structures and processes are needed to promote human well-being, as well as the issue of what is needed to promote and protect the well-being of the non-human, within the principles for conducting moral trade-offs already considered.

Given that, as we have seen, ecologism posits the need to establish guardianship institutions for the protection of the moral claims of the non-human, we might ask why ecologism should not similarly posit the case for such an institution to protect the claims of humans. This would in effect be to establish a body of Platonic guardians and would, of course, be a decidedly non-democratic form of government.

However, a political ecologist has reasons within the ecological outlook itself for resisting such a conclusion. In considering the well-being of any creature the ecological approach requires us to examine its characteristic capacities and relationships. In examining human beings in this way, ecologism will recognise that the crucial feature of the human species is that all its normally developed members possess the attributes of personhood.[2] This implies that, as persons, human beings have the capacity for reflection, for the acceptance of values and principles for the conduct of their lives, for self-criticism and for the joint creation with their fellows of the rich cultures which are their hallmark as a species.

Hence, the obvious reason for denying that human beings need to have a body of guardians to protect their moral standing and promote their well-being is that each human being, once it has reached maturity, has the capacity, other things being equal, for doing this itself. From this we may conclude that the most defensible method for looking after the well-being and moral status of individual human beings is to enable those beings as far as possible to rule themselves. Given that human beings need to reach decisions jointly on the best course of conduct, for this is in large part what it is to be culture-creators, then we have a case for joint self-rule, which is but another name for democracy. This is, in effect, the kind of case for democracy – and participatory democracy at that – which was put forward by Carol Gould.[3]

One reply which might be made to this raises fundamental questions about the character of ecologism as an ideology. The reply is that in developing this kind of argument ecologism is not putting forward a distinctively ecological approach, but is rather helping itself to the tradition of radical theorising about democracy which stems from Rousseau and is embodied in modern radical intellectuals such as Gould. Does this not bear out Martell's claim that, with regard to many issues of political philosophy, ecologism has no distinctive contribution to make, but has to turn to other, more traditional, political philosophies?[4]

To deal with this objection we must make a slight detour into the issue of the relations between ecologism and other politically oriented ideologies. Two responses may be made to the objection. First, the fact that two political philosophies share presuppositions and conceptions is no insuperable barrier to our regarding them as distinct. This is true, for example, of those offspring of the Enlightenment, liberalism and socialism. Secondly, we must further develop the idea that ecologism rejects the hitherto rigid distinction between the approaches of the natural and social sciences/humanities to the study of humanity.

The existence of recently developed biological sciences such as ecology gives us a new way of conceptualising the arguments and analyses of the great social and moral theorists, such as Rousseau. They can be seen as, in effect, carrying out a partial ecological analysis of the animal species *Homo sapiens.* The exploration of human nature and culture to be found in the great theorists is, in effect, human ecology *avant la lettre.* It has been partial because it has worked with only half of the picture, largely ignoring the fact that human beings are a species of animal, a matter which has been left to the physical and biological sciences to investigate. But an ecological study of this species of animal will need a full appreciation and understanding of its psychological capacities and needs and interrelations with its own kind as well as with other species and processes.

The two activities of social and political philosophy and social and political science are logically distinct but highly interconnected. A social scientific analysis of a reasoning creature is itself a specimen of what it aims to study. The reasoning produced by the creature as object of study will underly the theorising produced by that creature as that which carries out the study. The conclusions reached by such a creature *qua* philosopher about its own nature will inevitably be relevant to that creature's conclusions *qua* social scientist about itself.

It will be easier to appreciate these points if we note that an ecological study of our closest living relatives, the great apes, requires a similar attempt to understand them in terms of their social interactions. This in turn involves an analysis of, for example, their capacities for rudimentary language use, their creation and transmission of elements of culture, such as tools, and their capacities for self-awareness and forward planning. The recent studies by eminent primatologists such as Jane Goodall have been revelatory in these areas.[5] For a full ecological grasp of such creatures we need knowledge of such factors in addition to the knowledge of feeding patterns, reproductive rates, territoriality, and other behavioural manifestations; of the biological structures of the animals' bodies and metabolism; and of the animals' interactions with their habitats. Given all these kinds of knowledge we can begin to offer normative views about what form of primate society will be conducive to the well-being of such creatures.

We are just like such creatures with the crucial difference that we can study ourselves. This capacity for reflexivity is a boon, for we can speak of ourselves from the inside, and a drawback, for we are all too likely to have difficulty in understanding what precisely it is that we are up to, and to fail to see the mutual relevance of different kinds of approach to the problem of self-understanding.

Of course, any theorising about any subject-matter will be open to differences of opinion about where the truth lies. This applies to ecology as well. Thus the ecological argument for democracy outlined above, resting as it does upon a specific view about human capacities, is in principle as open to question within human ecology as it is within traditional political philosophising. But that does not undermine the case being here made for the view that ecologism has a distinctive approach to the issues of political philosophy. Different human ecologists may differ over the correct analysis of human nature, just as two primatologists may differ over the nature of a species of great ape. But what is distinctive about them as human ecologists, and what sets them apart from other normative theorists of politics, is that their views about the characteristic capacities of human beings are elements of a perspective which endeavours to grasp those beings in their total context, as a species of animal inhabiting a complex biosphere.

It will be readily apparent that what is going on in this argument is part of the familiar and normal struggle in which rival ideologies seek to give an account, in their own preferred terms, of the existence and appeal of their opponents. Thinkers whose primary allegiance is to one of the existing ideologies, even when they are sympathetic to views taken by ecologism to the issues which it raises, will seek to interpret its claims as essentially cautionary, requiring no more than comparatively minor amendments and extensions to the position of their preferred ideology.

By contrast, defenders of ecologism will seek, along the lines indicated in the last few paragraphs, to reinterpret their opponents as having at best a partial grasp of truths whose full force only emerges when they are set in the theoretical context provided by ecologism itself. This turning of the theoretical tables is something which ecologism cannot avoid. It is not a rhetorical device, but is inherent in the logic of the theoretical position of ecologism itself.

After this digression on the relations of ecologism to other ideologies, let us return to the matter of this section, which is ecologism's view of democracy. The general argument for democracy which ecologism supports and which derives from the human individual's capacity for self-rule, or autonomy, leaves many other more specific matters to be discussed, such as the crucial, interrelated, issues of which decision-making contexts require some form of democracy, which particular forms of democracy are appropriate in various such contexts, and whether ecologism is committed to a particular view of these matters.

An obvious issue to begin with is the claim that the acceptance of the autonomy argument for democracy is potentially in conflict with the requirements of ecological justice. That is, there is no guarantee that democratically organised human societies, whatever the model of democracy adopted, will respect the moral claims of non-humans for habitats and the wherewithal for flourishing, or even survival. Goodin has argued that Green political theory in general is oriented towards outcomes, particularly environmental protection, rather than processes, such as democratic decision-making, which cannot guarantee any specifically desired outcome.[6] This thought has seemed to some to commit Green political thought to at least countenancing authoritarian solutions to environmental problems, particularly given the pressure of time.[7] Democracy is too slow and too uncertain a way of arriving at decisions designed to protect the environment.

The force of this argument has in part been conceded in the last chapter, in which guardianship institutions have been accorded at least a limited power of veto in order to ensure that ecological justice is respected. It also seems requisite that the members of such institutions be at least partially shielded from the pressures of democratic politics by their holding office for a reasonably lengthy period of time. This point, together with the need for ecological and other forms of technical expertise, suggests that such a guardianship institution be appointed from suitably qualified candidates on a non-party basis by democratically elected governments for fixed, longish (say, ten years) terms of office. Removal from office prior to completion of the allotted period should only be for high crimes and misdemeanours. The justification for this limitation of the direct reach of democratically organised electorate is, of course, that the morally considerable interests of creatures unable themselves to engage in democratic politics need to be safeguarded by limiting the tendency of the exponents of such politics to ignore the interests of such creatures, even when the politics are conducted by ecologically aware citizens.

At this point it is appropriate to consider a suggestion put forward by Andrew Dobson in the context of developing an environmentally sensitive version of democracy.[8] He notes that there have come into view three 'constituencies' (already noticed by us in Chapter 5) of which traditional theories of democracy have been ignorant and whose interests it is now imperative to address – future generations, contemporary aliens and the non-human. All of these have legitimate interests which may be adversely affected by the environmentally damaging activities of given societies. He addresses the question of how these interests may receive direct (that is, non-paternalist) representation in the democratic legislatures of such societies.

In the case of future generations and non-human species he produces the ingenious suggestion that sections of the electorate in such societies should act as proxies for these otherwise unavoidably disenfranchised groups. They will elect representatives whose job it is to articulate and defend the interests of the groups in question. He envisages the electorate as comprising citizens who are already committed to the idea of environmentally sustainable development, and who will, accordingly, be willing to choose to abandon their normal franchise for the sake of the proxy one. The elected representatives will be accountable to the proxy electorate at election time for their conduct during the previous period of legislative activity. Genuine policy alternatives may be expected to be offered by rival candidates, given the fact of existing differences of opinion among the environmentally concerned about how best to protect environmental interests.

Dobson argues for the 'proxy' solution as being genuinely democratic – it secures accountability of representatives and is thus preferable to the paternalistic alternative of the appointment of individuals to represent indirectly the interests of the disenfranchised groups. This is a suggestion with considerable appeal and may well appear preferable to the paternalistic idea of the guardianship body mooted above. It is worth making the point that such a body is not totally divorced from democratic accountability, for its personnel is put in place by democratically elected politicians who, therefore, periodically get the chance to alter its composition in line with the views of the electorate, much in the way that the ethos of the US Supreme Court changes its ideological complexion over time in line with changes in the beliefs and agenda of the US electorate.

The choice between these two proposals for securing the representation of otherwise disenfranchised groups will turn on the feasibility of the 'proxy' idea. It requires a form of whole-hearted devotion to the interests of the groups in question which might not prove to be attainable, or attainable with any consistency, among even the most environmentally sensitive humans. These, after all, will still live in the human world in which such issues as jobs, education, defence and so on will still be of direct concern to them and their families. They may be very reluctant to give up their chance to affect these issues directly through their vote, and their orientation to the interests of the disenfranchised groups may be of uncertain quality. Arguably, such focus on the interests of future generations and the non-human will be easier to maintain among a group of people whose full-time job it is to devote themselves to these issues. Hence the idea of the guardianship institution, less than fully democratic though it may be, seems a preferable option for ecologism.

However, although ecologism is committed to this extent to a restriction of the reach of the full power of democracy, it does have good reasons to support democracy and to support its extension into areas where it is not always found at the moment, even within self-styled democratic systems.

Let us return, first, to the autonomy argument for democracy. Given the importance of autonomy for human flourishing, ecologism is able plausibly to claim that engaging in the processes of democracy is, at least in principle, part of the good life for human beings. Thus, as Dobson has also argued, Goodin's distinction between processes and outcomes should be rejected in this instance.[9] If the outcome you seek is the flourishing of human beings, a goal to which ecologism's general value-position strongly commits it, then you must institute democratic processes among such beings. Their autonomy, dignity, sense of self-worth and the full development of their social natures all require it.

Further, to adopt yet another argument of Dobson's,[10] ecologism's reliance upon the findings of the natural and social sciences also commits it to the general epistemological position that the best way to ensure the discovery of the truth concerning matters important for human and non-human flourishing is to ensure that the give and take of critical debate essential to, and fostered (at least in principle) by democracy, is permitted and encouraged.

Of all ideologies, ecologism, given the scientific basis of its factual claims, ought to be fallibilistic about such claims and encourage their scrutiny in a manner which democratic politics alone is capable of permitting. All ideologies, given the problematic nature of value claims, ought to be fallibilistic about these too, and this obviously applies no less to ecologism. There is a tension here, which is essentially the same for supporters of all ideologies, between the need to defend the values of the ideology to which you are committed and the need simultaneously to remain open to persuasion that it is in fact untenable. As Dobson notes, there is an act of faith in all this, central to scientific theorising as we have come to know it, which is that if what is said is true, then there is nothing to fear from its being critically scrutinised. If it is false, then even those who maintain its truth have an interest in discovering their error.

As far as the extension of democratic decision-making goes, the arguments for this will be better pursued when we turn to consider the issues of political economy in Part Four, for it is above all in the area of economic decision-making that ecologism sees the greatest threat to environmental well-being as turning on the absence of proper democratic processes.

Enough has been said at this juncture to establish that a commitment to democracy, albeit with a clear grasp of its dangers for ecological and environmental justice, is central to ecologism and not an optional add-on borrowed from other traditions where it is supposedly more at home. In discussing the case for democracy as put forward by ecologism we have also begun to address a matter raised under issue-area (6) in the last chapter, namely ecologism's account of its own self-correction processes.

Ecologism and Globalism

Do not the distinctive concerns of ecologism lead it inevitably to propose that human political structures must be global in scope? That is, is not ecologism an internationalist theory right from the very beginning? If not, why not? If so, what global political structures will it propose, and how will they work? In answering these questions we will need to consider such issues as whether human culture must inevitably be specific, bounded and contrasting with others in order to be viable for human beings; how democracy and accountability are to be achieved at any global level of government; how a global equivalent of the guardianship institution is to be maintained; how tyranny is to be prevented if world government is to be implemented.

Like its rivals, liberalism and socialism, ecologism is universalist in scope. That is, the aim of ecologism is to produce an analysis of the human condition and of moral considerations and constraints which are meant to be valid across the whole of humanity. Ecologism, based as it is in the scientific analyses of the human species provided by evolutionary theory, biology and ecology, offers a theory of human beings and the human condition which is meant to be rationally supported and true of human beings whatever their cultural formations.

Hence, while ecologism is able to find a place within its outlook for the explanation of the specific cultural formations of particular human groups, by relating them to the non-human context within which those groups live their lives, it is not thereby committed to the acceptance of all such cultural formations. For some of the latter, given the ecological analysis of the conditions of human well-being alluded to in the previous section, will have to be criticised and perhaps rejected as harmful to either or both humans and non-humans.

This implies that ecologism necessarily has a critical approach to human social and political arrangements and that it is willing to countenance only those local variations in human societies which are compatible with the universal requirements of ecologists' moral and political theory. This may still allow a wide range of possibilities for human cultural variations – in art, customs and traditions, forms of interpersonal relations, and so on. The precise determination of this will have to be taken a case at a time, of course. It does mean, however, that ecologism shares with liberalism and socialism the problem of how to convert to its point of view societies with totally different perspectives on the issues of human nature, the human condition and the requirements of morality.

This raises the fourth issue introduced at the start of this chapter, namely that of how ecologism should view the process of achieving agreement on matters of environmental importance with people who support other ideological positions, perhaps embedded in cultural traditions, such as religious world-views. Clearly, given its general moral position and commitment to ecological justice, it cannot simply adopt a policy of live and let live. On the other hand, it is committed by its view of the constituents of human well-being to a respect for the conditions necessary for the autonomy of human beings, which means having an attitude of prima facie respect for the traditions and cultures within which people seek to exercise that autonomy. We also saw in the last section the importance which ecologism essentially attaches to democracy, fallibilism and the give and take of rational debate as the sole reliable route to the truth.

Numerous exponents of environmental philosophy have put

forward a version of the ethics of discourse to seek to address this issue.[11] It is important to note two types of situation in which such an ethic may be held to apply. The first is when the debate in which the discourse takes place involves people who share a cultural formation and are not deeply divided by fundamental ideological differences. When the matter concerns deliberation on a practical matter on which they are all affected, the ethics of discourse requires that they respect each other's right to articulate their point of view, listen to the arguments carefully and seek to evaluate them rationally, openly and honestly admit clashes of personal interest and actively seek mutual accommodation in the general interest. Where no insurmountable moral dilemmas exist and all have an interest in reaching a joint decision then this ought to enhance the possibility of reaching an agreed outcome which all can see to be morally defensible.

However, this situation is very different from that in which the discourse is between individuals who are divided on matters of ideology and where the ideology in question forms part of a larger, more or less integrated cultural formation, within which the individuals concerned have reached their sense of self-identity. In such a debate the ethics of discourse outlined above may be followed assiduously and yet there may be no agreement possible because the commitment of the participants to their incompatible positions is so strong that no compromise is possible, and the idea that 'the best argument wins' has no real part to play, except perhaps in the very long run where fundamental positions come to be modified in the light of an ongoing process of critique and counter-critique.

There is still scope here for the idea that the basic respect which ecologism must accord to the conditions of human autonomy rules out the abandonment of discourse in favour of coercion, except in the most extreme cases of severe ecological injustice and shortage of time in which to secure it. But there is obviously no simple method or institutional arrangement which can preclude such a possibility from arising. Hence, the idea of an ethic of discourse has limited usefulness in such circumstances, except as an essential regulatory ideal.[12]

It is appropriate at this point to address briefly the highest-level theoretical issue in this whole problem area, which is posed by the argument that it is impossible to establish any foundations possessing certainty in any area of human thought, whether we are considering empirical or normative matters. Rather, it is claimed, we always find ourselves, as individuals and as societies, embedded in on-going structures, or quasi-structures, of thought in which webs of belief have been woven by human beings from within specific social contexts and sustained by specific social practices. Certain propositions may be treated within such contexts, for a period, as having a specially privileged position, often as the result of the exercise of social power by groups in whose interest it is for such propositions to be regarded as indubitable. But over time, with the processes of social change within which the interplay of discourse and power play a key role, different propositions and theories come to have a privileged position, as the web is extended and rewoven.

This is the kind of position, expounded in various forms by pragmatists, post-structuralists, Postmodernists, Foucauldians, communitarians and immanentists, which involves the denial of the possibility of any transcendent, neutral, indubitable, culture-transcending position upon which human beings may converge and from which they may reach agreement upon the empirical-cum-evaluative beliefs which they are to hold.[13]

Agreement, on the foundationless, immanentist view of these matters is achievable only by dialogue and negotiation, is always interim and never definitive, is always threatened by exercise of power on the part of those who wish to impose their own 'definitive' solution, and is in principle endless. There is no central truth to be discovered in such discursive practices, but only mutual accommodation in an on-going process of dialogue.

This view is to be distinguished from the superficially similar fallibilist position, which agrees that dialogue and mutual criticism are essential to the business of conducting the scientific and moral dialectic within which humanity is engaged. This fallibilist position is nevertheless committed to the view that there is a truth to be discovered, and there are more and less defensible value positions to be discerned. Criticism and debate must never be foreclosed, but the possibility that, over the long term, human beings are converging upon more adequate beliefs, must never be abandoned. Indeed, as critics of purely relativist positions never fail to point out, the exponents of the views expressed in the three paragraphs prior to this one must be taking themselves to have discovered the truth of the matter, which entails that there is a truth of the matter to be ascertained, contra to what is expressly claimed.

Ecologism is in the latter camp. It is committed, as we have seen on numerous occasions, to a science-oriented and naturalistic position, albeit a fallibilistic one too. It is universalist in its theoretical and moral positions and, as we noted in Chapter 3, has available a Darwinian position which appears to be able to explain the genetic, evolutionary basis of our theoretical and evaluative tendencies. For these reasons, ecologism is committed to the rejection of cultural relativism. It is willing to view the latter as making useful cautionary points about the ways in which inter-cultural dialogue should be conducted and as enjoining the valuable sense of humility which needs to be present in anyone who ventures to propose a viewpoint, whether empirical or normative, which purports to have universal validity.

It is also able to accept, again in a cautionary vein, one version of immanentism, namely that proposed by Walzer.[14] This claims that the way in which moral debate has to be conducted in order to have any practical power is always from within some moral tradition or other, embodied in the thoughts and practices of a specific society. It is possible, on this view, for representatives of different cultures to reach a transcendent level of discussion, but inevitably what one finds at that level are propositions of such generality and vagueness as to be practically inefficaceous. Nevertheless, on this view the transcendent level is admitted to exist. Hence, ecologism should draw from this view the lesson that, in order to optimise its chances of persuading adherents of cultures whose value-positions are problematic from the point of view of ecologism, the best plan is to seek to start from value-positions already accepted within those cultures and to draw out implications of those positions which point in the direction ecologism favours, and to criticise unfavourable value-positions on the basis of internal incoherence with those more congenial to ecologism's purposes.

There is no guarantee that this can be done, of course, in every instance. However, where it can be done it should be. There is no need for this approach to be regarded, by either proponents or critics of ecologism, as an attempt at manipulation or subterfuge. Proponents of ecologism can be quite open about what they are doing, for what they are doing is no more than what takes place within the natural evolution of cultures (at least within those which have not completely ossified), namely their self-transformation by internal critique.

Let us return to the main issue of this section, the inherent globalism of ecologism. It is important to notice that there is another basis upon which the inherently global character of ecologism is established. This, unlike the universalism just examined, clearly differentiates ecologism from its philosophical rivals. The difference may be highlighted as follows. Although it makes sense to suppose that a liberal or a socialist mode of socio-political organisation could come into existence within one society ('socialism in one country'), ecologism necessarily is committed to the implementation of its prescriptions across the whole globe, and to the achievement of this with immense rapidity.

Clearly, this is because of the nature of the values which it seeks to protect. In particular, the protection of biodiversity means that for ecologists the whole planet has to be the object of political concern – 'think globally', as the well-known Green slogan has it. An ecological dispensation in only one country will largely leave unrestricted the potential for the wholesale destruction of species and habitats by human activity – even if we are talking about a large country in a biodiversity 'hotspot'.[15]

If supporters of ecologism were given a free hand to devise a political system to protect their values and purposes they would, clearly, aim to operate with a global system right from the very start. This would undoubtedly involve the creation of more limited subsystems, perhaps based on bioregional considerations,[16] integrated into an overall gobal system of governance and structured in accordance with the requirements of democracy. There would be provision for the institution of guardianship for the interests of the non-human, as already outlined. The subsystems could allow for local cultural variety within the limits of what is required to supply the social basis for a life fit for human beings (as revealed by human ecology) and the protection of biodiversity.

However, supporters of ecologism are not in that position. Their problem is rather that of achieving something along the lines of this ideal blueprint in a human world already divided into separate polities on the basis of sheer happenstance, containing a diversity of human cultures, many of which are blind or hostile to ecologism's values, and which are often bolstered by a form of nationalism which at the very least tends towards the chauvinistic.

However, the fact that some countries are more strategically important from the point of view of biodiversity than are others suggests that, although ecologism's political perspective is inherently global – the biosphere is one system – some parts of that system are in practice of more pressing concern than others. Hence, the priorities for the protection of biodiversity suggest that ecologism should give its main consideration to the means whereby the 'hotspot' countries may be persuaded to adopt the main practical recommendations of ecologism. In practice, this amounts to the persuasion of tropical countries which contain the largest areas of rain forest to adopt conservation measures which work.

What this divergence between ideal and reality shows is that the clear difference in theory between ecologism on the one hand and liberalism and socialism on the other tends to diminish in practice. Just as the latter two ideologies have spread across the globe in a piecemeal fashion, so too must ecologism, even though, unlike the other two, ecologism is committed to the theoretical position that only a political dispensation which is global can provide an adequate response to the problems for which political systems are required. By contrast, liberalism and socialism can happily exist for extended periods of time, as they have in practice, at the sub-global level. This means, of course, that ecologism has no choice but to operate in practice with a sub-optimum set of aims.

This might be viewed as a cause for gloom, but it can be given a more optimistic cast if we notice the importance within the contemporary world of two important phenomena. One is the success which liberalism, especially economic liberalism, is having in becoming a genuinely global phenomenon. The other is the rapid spread of international mass communications, especially via the internet. These two 'globalising' phenomena are helpful to the cause of ecologism in various practical ways.

For one thing, the spread of the moral ideals of liberalism in connection with human beings, such as the concepts of human rights, autonomy and self-development, creates the possibility of a unitary moral vocabulary within which to spread the moral ideas of ecologism. Once people have accepted the moral ideas of liberalism it becomes possible to mount the 'consistency' arguments mentioned in Chapter 4. Ecologism has sprung from cultures in which the moral ideas of liberalism have been dominant, even if ecologism represents a significant challenge to those ideas. This historical point arguably has a more fundamental importance – that an acceptance of liberal moral notions with respect to human beings is a facilitating step towards the acceptance of ecologism as a moral theory. This point is an adjunct to the point made above with respect to immanentism, namely that proponents of ecologism should also seek to build arguments towards its conclusions on the basis of any promising value-positions there may be within the traditions of specific communities.[17]

In addition, the spread of political and economic liberalism also facilitates the spread of scientific concepts of, and approaches to the study of, both the natural world and human societies. This allows the spread of the concepts essential to the grasp of ecologism. The interest all societies have in agriculture and husbandry, climate and the control of destructive natural forces, the realisation of biological resources (as in biotechnology), provides the motive for people within such societies to gain the mastery of such scientific concepts and theories.

Once these become pervasive then the fully contextual nature of human beings in the biosphere becomes easier to grasp and accept, and with it ecologism's moral and political project. This is not all plain sailing, by any means. It is clear that older, unhelpful attitudes to the non-human world can coexist within the same mind with scientific understanding.[18] This is as true of Western as of non-Western minds. Thus, enlightenment is not guaranteed. The most that can be said is that the spread of scientific understanding provides an opportunity for ecologism's claims to be expressed and to begin to take hold.

As Low and Gleeson have noted,[19] the growth of instantaneous global communications networks also provides the means for the transmission of ideas supportive of ecologism's position, although the channels in question are undoubtedly going to be cluttered with viewpoints, issues and data which may be harmful to it. Of course, the internet, impressive as its growth has been, is still not going to reach everyone whose opinion it is important to affect. But it is undoubtedly another powerful agent for the creation of a set of globally understood concepts within which ecologism can be articulated and made known – even if only to elite groups in certain countries.

Another phenomenon which gives ecologism some cause for hope that its globalist orientation is not entirely barren is that there do already exist elements of a world government in the development of the UN and its associated institutions as well as an ever-expanding body of international law and international regimes at least some of which promise to be supportive of ecologism's aims. Much remains to be done to turn such elements in the direction of planetary protection, and there are arguably more reasons for pessimism than for optimism in the way global processes of governance are developing. We are also some way away from viable institutions of world government, although we seem to have already in place a variety of structures and practices which embody a recognisable form of world governance.[20]

Ecologism and States

Under this heading we start to encounter more directly the elements proposed under issue-area (5) from the last chapter, namely the attempt to integrate the findings of political science into the discussion of political philosophy so as to refine the concepts of what is feasible in human political arrangements and to determine where various sorts of danger lie.

The issue of the role of the state, both in theory and in the contemporary world, has been vigorously debated in recent years from a variety of perspectives. Theorists of the public choice school have sought to analyse the ways in which the existence of a state apparatus enables those who occupy the governmental roles thereby created to pursue their own interests, in close collaboration with powerful interest groups.[21] On this view the modern, extensive, interventionist state inevitably becomes a conspiracy against the public weal. The way to deal with this threat, these theorists argue, is to restrict the state to something like the minimalist role of classical liberal theory.

Specifically with regard to the role of state ownership of resources, especially land, theorists who defend the regime of private property argue that only when property is in private hands is there the motivation to look after it properly and to safeguard the natural resources, including non-human creatures, which it may contain. Hence, on this view, the solution to such environmental problems as the tragedy of the commons is not to establish a coercive state which, in the pursuit of the common good, forces individuals to restrain their self-interest, for such coercive state power is always captured by powerful groups and used for their narrow interests. The solution is to convert the global commons, including air and water, into private property which its owners can use to pursue their private interest in the market-place, thereby enabling the 'invisible hand' celebrated by Adam Smith to work its wonders.[22]

Carter has offered a version of these anti-state views from within the environmentalist camp, seeking to show that the modern state is inevitably locked into an environmentally pernicious dynamic, in accordance with which it is in the self-interest of state officials, whose power and wealth depend upon it, to encourage an aggressive form of capitalism in which the production and sale of armaments play a central role, justified by arguments of national security.[23] To break out of this into an environmentally beneficent dynamic the modern state has to be systematically replaced with decentralised, low-tech, self-sufficient communes which directly intercommunicate so as to defuse the potential sources of international hostility.

The minimal-state/anarchist tradition is clearly at work here. In terms of the traditional liberal distinction between the state and civil society, these critiques of the state posit the need to rest a great deal more on the workings of civil society structures – whether these are conceived of in terms of the economic market, private property or self-governing communes – than has been characteristic of the modern state.

Defenders of the state regard these critiques as partial and overly sanguine about the possibility of a satisfactory form of political organisation outside the state. The public choice and private-property theorists face the criticism that their proposals are a recipe for power and wealth to flow into the hands of those who succeed in dominating others in the market-place. Greater openness, greater accountability, greater devolution of decision-making to the lowest practicable level are seen as the best ways to counter the misuses of state power identified by the critics. An example of what this might mean can be found in Dryzek's discussion of the development, usually instigated by states, of what he calls the 'public sphere'. Within this there lies, he suggests, the possibility for ordinary citizens to develop those practices definitive of discourse ethics which are usually so clearly absent in the workings of central state institutions. Such practices as public inquiries and mediation forums are examples of what he has in mind.[24]

The state is seen by its defenders as needed in order to play a key role in regulating abuses of private, especially financial, power, particularly that associated with multinational corporations; to play the crucial role in preventing the tragedy of the commons; and to play another crucial role in co-ordinating decision-making among lower-level units, thereby ensuring that decisions by one sub-group do not stymie those made by another.[25]

The position on the role of the state taken by ecologism will depend on various factors. Let us remember to begin with the distinction which emerged in the previous section between the ideal theory of ecologism and the practical political strategy which ecologism has to countenance in the world as it actually is. In the area of ideal theory is there any distinctively ecological reason for favouring state-like modes of political organisation? Given a blank sheet, would ecologism inscribe the state upon it?

The answer to this depends in part on what we are to suppose the basis of delineating particular states ought to be. Are we to suppose that each state is to be a nation-state, so that the territory and membership of the state is determined by membership and location of a specific national group? If so, then the justification for this is going to require the claim that, by and large, human beings are grouped into such nationalities and that their membership of these is one of their vital interests, essential to their well-being.

We may, however, be invited to consider the formation of states, each claiming the sole legitimate right to use coercion within a given territory, on the basis of bioregions. This would then be justified on the basis that from ecologism's point of view, the goal of politics is to protect the interests of both human and non-human inhabitants of this planet in the most efficacious way. This would need to be coupled with the claim that the protection of the human and non-human within what is clearly a unified biosphere can best be achieved if political boundaries are drawn in such a way as to respect the natural boundaries of ecosystems within which life-forms flourish. This assumes that specific ecosystems can in practice be delineated in such a way as to give a clear and ecologically defensible compartmentalisation of the Earth's surface upon which to base political structures. This may not in fact be possible, or be too indeterminate to be anything but arbitrary.

Clearly, these alternatives are likely to give different results. Human beings, as is characteristic of their species, occupy the whole variety of ecosystems available on the planet. A single human nation, on the first criterion, may occupy a territory which straddles several bioregions, supposing we can satisfactorily delimit these. Even the relation of human cultures to the ecological substrate will not necessarily be helpful here – there is no reason why one and the same national group may not contain various subcultures.

The problem here derives from the fact that human cultures are not always closely connected with underlying ecosystems. Although there may be some rough correlation, there is a lot of free play too between human groups and the structure of their natural contexts.

In part this is due to the fact that human groups acquire elements of culture from other human groups whose culture developed in very different ecosystems; in part it is due too to the fact that as human beings progressively build around them an artifactual world the cultural interconnection is progressively loosened between them and the ecosystems of the territory which they inhabit.

There is reason to suppose that the sense of membership of a culturally determinate and differentiated national grouping is necessary for the self-identity of human beings.[26] Thus, there is some reason for a human ecological approach to favour giving priority to the existence of distinct human cultural groupings in considering an argument for creating states. The indeterminacy of ecosystems may actually be helpful to ecologism at this point, since it removes any strong reason to give priority to the attempt to delineate states on the basis of ecosystem boundaries. A justification congenial to ecologism for basing states on human nations is that, just as territorial animals defend their territories and have an interest in their well-being, so too might human nationalities whose distinctive national cultures are at least in part linked to the ecosystems and non-human creatures with which they share a specific region of the planet.

However, even if ecologism can find ecological reasons for drawing up the borders of states on the basis of nationality, we still need an account of ecologism's reasons for having states at all. We saw in the last chapter the way in which the 'tragedy of the commons' could be given a foundational role by ecologism in establishing the case for a political order of some sort. But does this order have to be a state-like entity?

Further, where does ecologism stand on the traditional political philosophical arguments for the creation of states – such as the need to preserve human beings from internal depredation and external attack? What view does ecologism take on such issues as the need for social co-ordination and the possible role of civil society?

The need for protection against both neighbours and strangers has usually been defended on the basis of views about the essentially self-interested character of human nature and its limited capacities for altruism. Associated with these are human propensities to deal harshly with those whom we believe have wronged us and to ignore or trample on the interests of other human beings when a benefit to ourselves and our loved ones may thereby be secured. On the issue of how far these views represent the truth about human nature the approach of ecologism will be to try to ground any such views on the scientific basis of human ecology as far as possible. Here considerations arising from evolutionary biology, sociobiology, psychology, primatology and so on will be relevant.

On the specific issue of whether we need states to protect members of human groups against internal malfeasance and external aggression, there is some reason to suppose that ecologism should side with the idea that such a tendency to aggression and selfishness is an ineradicable part of human nature, although the tendency to co-operation must be given its due too. There are good biological reasons for supposing this balance of tendencies to be present in human beings. Comparison with our closest relatives, the chimpanzees, suggests that such propensities are a feature of the genus of which we are a part. That being so, the view of ecologism will be that the traditional defence of the need to form state-like modes of political organisation is sound.

On the questions raised by the traditional modes of political philosophy, then, ecologism can find some reason from its own approach for arguing that, although its ideal perspective on political organisation is globalist, it can endorse the creation of sub-global states based on distinct nationalities and justified by the need to protect such populations against internal and external aggression.

Other standard arguments in favour of a sub-global level of political organisation which ecologism can accept are, first, that the compartmentalisation of the world in this way provides some safeguard against the possibility of a world-wide tyranny which might result if the global level of government were to fall into the wrong hands. This is also one of the key arguments in favour of a federal system of government at the level of the nation-state. Secondly, there is also the argument from practicality, namely that a global level of government will be unable to govern directly the whole world from the centre. It may have an overseeing and co-ordinating role, but detailed decision-making will inevitably have to be delegated to sub-governments more directly in contact with developments in specific areas. Thirdly, the practical implementation of democracy and participation across the globe clearly requires levels of political organisation lower than the global.

How does ecologism respond to the criticisms of the nation-state offered by the public choice school and the defenders of private property regimes? More generally, does ecologism have any reason to accept anarchist ideas to the effect that co-ordination of human efforts in the economy and in safeguarding the interests of the non-human can be best achieved on the basis of spontaneous order and forms of co-operation, rather than on some overt system of centralised planning?

The criticisms which have been offered of anarchist political concepts seem to be as strong as they ever were. Although there are nowadays quite sophisticated defences of the emergence of co-operation among human groups based upon games theory, as in the work of Michael Taylor, there are no clear examples of this working in practice among real human beings for any length of time.[27] There is no reason for ecologism to be hostile in principle to anarchist ideas about political organisation, but there is reason for caution in lending support to the attempt to put such ideas into practice, particularly when we are considering, as ecologism perforce has to, the issue of how co-ordination is to be achieved across millions of culturally diverse people living in a rapidly changing economic and social environment.

The case of the market economy is rather different, although here too there are enough reasons to doubt that the 'free' market works in the way its defenders suppose to justify scepticism of its ability to deliver such goods as ecological justice.[28] This is not the same as the question of whether biodiversity may not best be preserved by harnessing human economic motivation to the defence of species, by, for example, encouraging people in a specific locality to manage, and gain an economic return from, the endangered species with which they live in close proximity.[29] However, aims such as ecological justice appear only to be practically attainable if the economic activities of markets are carefully circumscribed by a legislative framework put in place, and maintained, by political bodies.

This point applies also to those solutions to the tragedy of the commons, or spontaneous disorder, which rely on the conversion of commons into private property regimes. There is no reason to suppose that the owners of private property will have the requisite motivation, deriving from the economic benefit they stand to gain from their ownership of such property, to pursue such aims as the preservation of habitats and biodiversity, as required by ecological justice. There is in any case the more general problem of the co-ordination of decisions across the environment which private property regimes on their own do not appear able to deal with. For example, two individuals who each separately own the two remaining specimens of a habitat may allow the habitat which they own to be destroyed in the false belief that the other specimen of that habitat will be preserved by the other owner.

To get over such problems an *ad hoc* co-ordinating body may emerge. But there is no guarantee that it will be effective in getting its efforts accepted by the participants and it may have only patchy coverage (it will, after all, cost money to create and maintain). These difficulties point in the direction of a body with the necessary power and resources to maintain co-ordination, and to attend carefully to the demands of ecological justice. It is plain that such a body has to be a political structure forming part of the state apparatus. The securing of justice has long been recognised to be a public good, and it is also accepted as the business of public bodies, such as governments, to secure such goods.

These kinds of issue, and the arguments on each side of the debates, have been discussed by theorists other than ecologists for purposes unconnected with environmental concerns. But there is no reason to suppose that that is a reason for saying that ecologists have no specific points to add to the discussion, and that thus ecologism can only latch on to existing political views, rather than offering its own distinctive perspective. It is clear that ecologism's specific concerns for ecological justice, involving the protection of habitat and preservation of biodiversity, do bring in specific reasons for criticising public choice and private property conceptions. These derive from the global and interconnected nature of the issues. If the biosphere could be divided up into discrete parcels and allocated as private property then the private property regime would have more to be said for it. But in the environment we have the paradigm of the interconnected.[30] We therefore need mechanisms for paying due attention to those interconnections and ensuring co-ordination of human efforts across the whole biosphere in pursuit of ecological justice.

We have seen in this chapter that ecologism has a distinctive approach to fundamental matters of political philosophy with respect to the large-scale matters of foundational considerations, democracy, the state and globalism. However, it may still be claimed that in relation to more specific matters of political philosophy, especially those concerning liberty, equality and justice within the state, ecologism has to turn to traditional theories of political philosophy for decision, having no distinctive position of its own to put forward. To determine the correctness of this claim we will have next to consider the state of the contemporary debate with respect to these issues, and examine what distinctive position with respect to them, if any, is to be found within ecologism.

Notes

1. See Arran Gare, *Postmodernism and the Environmental Crisis* (1995) for an enlightening attempt to bring the science-oriented discourse of environmentalism and the literature-oriented discourse of Postmodernism into fruitful conjunction. Roughly, 'Postmodernism' refers to a situation in which there is held to be a widespread loss of the sense that there is an overall progressive direction to human history. This sense of 'progress' once privileged some forms of knowledge and cultures over others. That is held to be no longer a credible position.
2. I further discuss the concept of personhood and its importance for ecologism in 'Ecocentrism and persons' (1996).
3. Carol Gould, *Rethinking Democracy* (1990).
4. Luke Martell, *Ecology and Society* (1994), p. 139.
5. See, for example, Jane Goodall, *In the Shadow of Man* (1971).
6. Robert Goodin, *Green Political Theory* (1992). This is a view which he has subsequently amended – see his 'Enfranchising the Earth, and its alternatives' (1996).
7. See, for example, William Ophuls, *Ecology and the Politics of Scarcity* (1977).
8. Andrew Dobson, 'Representative democracy and the environment' (1996b).
9. See Andrew Dobson, 'Democratising green theory'(1996a), p. 140.
10. Dobson 1996a, p. 139.
11. See, for example, John Dryzek, *Discursive Democracy* (1990).
12. Nicholas Low and Brendan Gleeson, *Justice, Society and Nature* (1998), chapter 8, address this issue and posit four principles of 'political justice' – roughly, the principles to govern political debate – at the global level. These are lexically ordered and comprise the ecological principle of justice, the principle of respect for human autonomy, the principle of uncoerced discussion and the principle of consent.

    There is, as I have indicated, reason to suppose that a concern for justice, ecological or environmental, cannot preclude the threat and use of coercion to secure such justice against recalcitrant actors in a situation where a serious injustice will be done in the near future, that is when time for debate is rapidly running out. The ethics of discourse and the institution of discursive democracy are regulatory ideals. In practice they may not be attainable to a full, or a significant, degree. Justice is not a negligible demand, and coercion to achieve it cannot be precluded in the global, environmental context any more than it can in the domestic, social one. Violent coercion should certainly be a very last resort, only employed to the minimum extent necessary and with a full awareness of the serious drawbacks it inevitably contains from both a moral and practical point of view.

13. See Gare 1995, chapters 1 and 2 for a clear exposition of the varieties of Postmodernism.
14. Michael Walzer, *Spheres of Justice* (1983), p. xiv.
15. The idea of such 'hotspots' is explained in E. O. Wilson, *The Diversity of Life* (1992), pp. 247–60.
16. The idea of bioregionalism is championed by Kirkpatrick Sale in *Dwellers in the Land: The Bioregional Vision* (1985).
17. This discussion of the possible benefits to ecologism of the spread of liberal ideas does, of course, run the risk of being regarded as another form of cultural imperialism. However, it is very difficult even to state this view without employing ideas central to the liberal tradition, such as respect for individuals' values and beliefs, the right to autonomy, democratic debate and agreement based on discussion, not force. It also presupposes a reification of cultures which arguably is not sustainable.
18. See Richard Harris, 'Approaches to conserving valuable wildlife in China' (1996), p. 304 for an example of how a trained Chinese zoologist retained the traditional Chinese belief that 'wolves are bad animals'.
19. Low and Gleeson 1998, p. 204.
20. This distinction between governance and government at the global level is taken from Low and Gleeson 1998, p. 185.
21. See, for example, William Niskanen, *Bureaucracy and Representative Government* (1971).
22. See John Dryzek, *The Politics of the Earth* (1997), pp. 102–8 for a discussion of the exponents of private property regimes.
23. See Alan Carter, 'Towards a green political theory' (1993).
24. Dryzek 1990.
25. See Marius de Geus, 'The ecological restructuring of the state' (1996).
26. This is a claim characteristic of conservative and communitarian philosophers. The use of the claim made by the latter is discussed in Chapter 9.
27. See Michael Taylor, *The Possibility of Cooperation* (1987).
28. See Dryzek 1997, chapters 3 and 6, for an excellent discussion of the problems.
29. The development of the 'Campfire' project in Zimbabwe appears to be a successful case in point. See *The Guardian*, section G2, 20 May 1997 for a report on this by Liz McGregor.
30. Dryzek 1997, chapter 6 is, once again, eloquent on such points.

8

Ecologism and Contemporary Political Philosophy: Utilitarianism, Rawlsian Liberalism and Libertarianism

IN THE PREVIOUS CHAPTER we considered some of the large-scale issues of political philosophy which involve a distinctive approach on the part of ecologism. However, a great many issues broached by contemporary political philosophy require a more closely focused discussion of the internal socio-political arrangements of human societies than we have so far given. In this chapter, therefore, we will look at some of the more specific issues, largely involving a focus on the domestic rather than the international arena, which have formed the subject of debate in political philosophy since the 1970s so as to indicate the distinctive views of ecologism in these areas.

In order to do this systematically and in a way which will be helpful to those familiar only with mainstream political philosophy, I will follow the sequence of topics in Kymlicka's excellent recent survey of the subject.[1] It is notable that in that book the environment receives only a couple of tangential mentions, which indicates from the outset how lamentably partial and inadequate recent political philosophy is from the point of view of ecologism, sophisticated though it is in other respects.

We need first to note the illuminating thread which runs through Kymlicka's discussion, namely the claim that the political philosophies which compete with each other for our allegiance are all committed to the same ultimate principle, namely that, in Dworkin's formulation, 'each person matters equally'.[2] On this view the influential philosophies of recent times – utilitarianism, Rawlsian and Dworkinian liberalism, libertarianism, Marxism, communitarianism and feminism – all attempt to explain what principles, practices and institutional arrangements are needed to do justice to this ultimate value.

This is a convincing view of the fundamentally egalitarian purpose of all recent theorising, even of theories, such as libertarianism, which on a superficial reading seem to downgrade many specific egalitarian concerns in favour of various forms of liberty. It is a view, therefore, with which we do not need to take issue. As we have seen, of course, what ecologism seizes on immediately is the purely anthropocentric view expressed in the above statement of the ultimate value, assuming, as everyone does, that only members of the species *Homo sapiens* are persons.

Utilitarianism, Consequentialism and Ecologism

What is the general character of ecologism as a moral theory? Is it like classical utilitarianism, concerned to promote certain overall states of affairs, such as the greatest happiness of the greatest number? Or is it merely in some sense consequentialist, concerned with the outcomes for the well-being of life-forms?

Ecologism is well advised to distance itself from classical utilitarianism, plagued as that is by well-known difficulties. Among these is the problem that utilitarianism is incapable of giving a satisfactory account of the particular commitments which people have to particular others – spouse, parents, creditors – which, as Kymlicka says, 'form the focal point of our lives and give some identity to our existence'.[3] This problem stems from the fact that utilitarianism views people as merely the means to achieving its ultimate aim, which is the creation of a state of affairs – the greatest happiness of the greatest number (of human beings). This has the effect of reducing all moral judgement to a single calculus in which individual human beings are receptacles of the desired property – happiness, pleasure or want-satisfaction. No particular projects or purposes of the kind which give individual lives happiness or meaning are given any particular weight. This, as Kymlicka argues, is to convert utilitarianism from a moral theory to something more akin to an aesthetic one. The overall state of affairs aimed for has lost contact with the well-being of individuals, which is, for a maximally individualised species such as human beings, the essence of morality.

Can ecologism escape similar charges? Can it give an account of the role of particular commitments in individual human lives?[4] Or does it, like utilitarianism, regard moral agents merely as means to the fulfilling of some desired overall state of affairs, such as the maximum preservation of biodiversity?

It is clear that ecologism is not committed to a 'flattening' of the moral life in the manner of utilitarianism. It operates instead with a notion of what is owed to morally considerable entities in order to permit them to attain the well-being appropriate to their natures. In the case of human beings, supporters of ecologism have every reason to accept that personal commitments and the associated exercise of autonomy on the part of highly individuated beings are essential to their well-being. The aim to preserve biodiversity is the aim to respect similar considerations with respect to other morally considerable entities.

As noted in Chapter 4, the argument that species and other non-human biological entities are intrinsically valuable because they are wonderful does employ a concept which straddles the moral/aesthetic divide. However, this argument does not commit us to project of maximising wonderfulness, which would land ecologism in the same predicament as utilitarianism. It rather gives us a reason for according moral considerability to entities which are not moral agents and so lack the usual basis for such considerability.

We have already emphasised that the property of moral considerability admits of degree, allowing discrimination and trade-offs to be made between the interests of beings with different amounts of it. Plainly, the particular commitments of specific human beings will often be of crucial importance for their interests and therefore such commitments have to be considered in making the trade-offs. However, ecologism will want to exercise the same caution with respect to human commitments which Kymlicka and others exercise with respect to human preferences in the course of their consideration of the 'preference-satisfaction' version of utilitarianism. The important point with respect to the latter is that only some preferences of human beings have legitimate moral weight. Notoriously, a human racist's preferences for poorer treatment for members of other races should not be counted when we are seeking to satisfy preferences.

Similarly, human commitments to others are only legitimately countable from the point of view of ecologism if they do not essentially involve ignoring the claims to moral considerability of other life-forms. If human beings find themselves committed to forms of life and/or personal relationships which lead them in serious ways to ignore such interests of other life-forms, then it is morally legitimate to require them to change their ways, albeit with help and support from their fellows. The end of the slave trade quite properly ended a whole host of commitments of many human beings, putting an end to some lucrative forms of life. So too will the ban on hunting and capture of other species, with equal moral justification. As Kymlicka expresses the point, 'Part of what it means to show equal consideration for others is taking into account what rightfully belongs to them in deciding one's own goals in life'.[5]

This, of course, raises the question of whether we can specify what rightfully belongs to non-humans in the course of determining how to treat them as possessing moral considerability – in other words, what is required by ecological justice. A general answer to this can be given for many such beings, given that they are usually not extensive environment modifiers, have evolved to occupy specific niches in the physical environment, do not have rich, changing cultures and so on. What they need overwhelmingly for their well-being, and thus what is rightfully theirs from the moral point of view, is enough habitat to be able to flourish and reproduce their kind in viable numbers. How much this is, and how strong a case they have, will depend on specific facts about the species in question. For this we need the expertise of field ecologists and others.

Thus, once again, ecologism is able to demonstrate the way in which considerations which arise in the case of moral interaction between human beings are simply more specific forms of more general considerations which apply between human beings and other species.

To return to the question with which this subsection began, it is clear that ecologism is consequentialist in the sense endorsed by Kymlicka.[6] That is, ecologism espouses the 'intuitions' that

(1) the well-being of all life-forms matters
(2) moral rules must be tested for their consequences on the welfare of life-forms.

Rawlsian Themes: Justice and Moral Considerability

We have already to a certain degree discussed issues of justice from the point of view of ecologism. We now need to bring the ideas already introduced into direct contact with influential recent discussions of justice, of which Rawls' are the most interesting.

Rawls' theory of justice as fairness applies only to human beings. It attempts to determine which rights and liberties must be distributed to human beings in a social arrangement which is going to merit the appellation 'just'.[7] This means, on the basis of the egalitarian principle noted in the last section, that the society in question treats persons as counting equally, morally speaking.[8]

Once again, ecologism requires that this issue be reformulated in line with its own ultimate value postulate. It requires that an answer be given to the question of what distribution of rights and liberties between human beings and between them and non-humans is needed in order to treat human persons as morally equal to each other and both humans and non-humans as being morally considerable. An answer to this question will provide a theory of justice which will govern both inter-human relations within human societies and the relations between those societies and the non-human realm within which those societies actually exist – a realm which theories such as Rawls' simply ignore. We are back again with the topic of ecological justice introduced in Chapter 6.

Ecologism also endorses the Rawlsian claim that 'justice is the first virtue of institutions', so that injustice cannot be redeemed by appeal to other values.[9] Rather, as Kymlicka expresses the point 'the legitimate weight attached to these ... values is established by their location within the best theory of justice'.[10] Since, for ecologism, human political arrangements have to answer to the moral claims of both the human and the non-human, the location of moral values within the 'best theory of justice' will be more complicated than it appears even in Rawls' elaborate theory. But it is an implication of ecologism's acceptance of the Rawlsian dictum just noted that a political system which is unjust, inadvertently or by calculation, towards the non-human cannot be redeemed by its furtherance of other values for humans or non-humans.

The actual theory of justice which Rawls produces in answer to the question he asks is intended to be, to use Dworkin's formulation, 'endowment insensitive and ambition sensitive'.[11] In other words, it is supposed to permit human beings the opportunity to acquire goods only on the basis of the personal efforts to which their ambitions move them, while excluding their acquisition of goods on the basis of features of their situation which are morally arbitrary. For Rawls, these morally arbitrary features encompass social position, family background and genetically determined features such as race, sex and talents and aptitudes. People are not treated justly if they acquire holdings on the basis of their possession of one or other of these characteristics.

Thus, they may only properly acquire holdings of goods on the basis of such a morally arbitrary feature as their genetic endowment of talents if they accept that they will be required, on the basis of what is needed to secure justice, to participate in a redistributive scheme intended to compensate those who have not been as well endowed as they are. Any other arrangement will distribute life-chances in a manner which does not respect human beings' equality of moral status. The differential rewards to which the more talented may justly aspire must, in turn, only be allocated on the basis of the actual mobilisation of their talents as a result of their personal ambition and preparedness to exercise effort to develop those talents. In other words, they must be open to what Rawls calls 'fair equality of opportunity'.[12]

We may not, then, discriminate morally between human beings on the basis of a 'natural aristocracy' of birth. Can this thought be turned by ecologists in a direction of a theory of justice which encompasses non-human creatures? There are various problems with this. First, social justice applies to human societies, conceived of as being, in Rawls' formulation, 'cooperative ventures for mutual advantage'.[13] Non-humans do not and cannot form part of such venture. Some of them may be brought into the venture, but not as a result of their freely choosing to co-operate. Thus the Rawlsian formulation of the issue of justice – to find principles of co-operation within such a venture which pay due respect to the equal moral standing of all the participants – cannot be used by ecologists to encompass the non-human.

It might be thought that the argument for compensation based on the moral arbitrariness of genetic endowment should be extended to non-human non-persons. It is, after all, a matter of genetic luck that human beings have been endowed with the capacities for personhood, whereas other creatures have not. Thus, surely on the basis of Rawlsian reasoning we should say that any advantage accruing to human beings from this fact should be viewed as undeserved and subject to the provision of compensation to those disadvantaged in this respect, namely other life-forms devoid of personhood?

However, this does not work as an argument. It makes sense to suppose, of given human beings, that they might have been born with greater or lesser or different talents from the ones they actually have, and therefore to conclude that the ones they actually have are a matter of genetic lottery. But to suppose that a parrot might have been born as a person is to suppose that it might have been born as a member of a different species. It is not, however, coherent to suppose that something might have been born a member of a different species from the one in which it was actually born, for what could that 'something' possibly be?

For this reason, too, the Rawlsian device of the original position could not be adapted by assigning proxies to represent non-humans devoid of personhood in the debate over basic principles of justice. That imaginary debate is conducted behind a veil of ignorance. But, for the reason just given, the device of the veil cannot coherently involve imagining the participants as being ignorant of the species of which they are members.

What these points suggest, then, is that a theory of justice such as Rawls' is specifically tailored to the issues of how matters of joint activity are to be conducted among persons (who happen, on this planet, to be confined to one species). Taken as such, there is no reason why ecologists should not endorse the basic ideas of the theory, which do seem to pay proper regard to what is needed to give appropriate recognition to the moral considerability of persons. Their capacity for autonomy is given due scope while preserving fairness of treatment between persons who are subject to morally arbitrary differences in initial life-chances.

Non-human beings, however, also have endowments and a life to lead in which their opportunities for flourishing merit moral consideration. Moral agents do not have to be concerned to devise principles which are ambition-sensitive to govern their relations with such creatures, for non-humans do not have ambitions. The endowments of non-humans, however, are a matter of legitimate concern, in two ways. First, the endowments of capacities specific to each species establish its conditions of flourishing, and the adequate recognition of the moral considerability of such species has to attend to those conditions and their maintenance. Secondly, the fact that non-humans, on this planet at any rate, are not persons means that they possess no capacities for articulating and defending their own moral considerability. This is a matter of their lacking an endowment. It is not arbitrary that this is so, as we have just noted, for it is partially definitive of their specieshood. But it does put them at a severe disadvantage in the matter of their interactions with persons.

There is, therefore, an issue of justice between species which are persons and species which are not which is a more generalised version of the issue of social justice between members of species which are persons and form societies. In fact, there is a series of issues here, of which Rawls (and most traditional political philosophy) has selected only one. For, as we noted in Chapter 5, there are issues of justice between persons who are members of different societies and between persons living at different times. There is also the issue of justice between members of different species who are persons. This is not, of course, a live issue at the moment, but it raises complications concerning the balance to be struck between persons with different kinds of needs arising from their different species characteristics, such as their modes of feeding and reproducing. There are also issues, discussed by Benton,[14] concerning moral connections between human persons and those other species which we have forced to join our societies and play key social and economic roles within them – creatures in our societies but not of them.

What is characteristic of the ecological approach to these issues is the basic concern to discover the conditions of flourishing of each such species and to derive the principles governing the due recognition of such species' moral considerability from those conditions. As we have seen in Chapter 5, the principles of social justice, which concern the distribution of resources (or resource-equivalents) among human beings so as to permit basic needs for all of them to be met, have priority over principles governing the distribution of resources between humans and non-humans so as to meet basic needs of the latter. However, it will be recalled from that discussion that what is needed to enable non-humans to meet basic needs can trump considerations based on what human beings seek to have to meet their non-basic (instrumental) needs.

Where basic needs clash in this way, however, the requirements of ecological justice put limits on the way in which human basic needs are to be met. The following constraints operate:

(1) when social justice requires the mobilisation of natural resources to meet human basic needs, human beings must seek ways of doing this which allow non-humans as far as possible to retain the conditions necessary to meet their own basic needs – that is, the moral permission to intervene in nature to meet human needs is not a licence to disregard recklessly the moral considerability of the non-human;

(2) if such basic human needs can be met either by mobilising natural resources, or by redistributing among human beings, then the latter should be chosen in those circumstances where the former destroys the conditions necessary for non-humans to meet their basic needs;
(3) human beings must seek to arrive at some conception of 'enough' to define the degree of material well-being and consumption which is acceptable for all human beings, so that the resources needed by non-human nature are not eroded by an open-ended increase in the level of material consumption counted as acceptable – this permits such increases only where human ingenuity can provide them without such erosion;
(4) even when the requirements set out in (3) have been met, human beings have the responsibility to limit their numbers so that they do not get into the position of having no alternative, in order to meet the demands of social justice, but to destroy the conditions necessary to the existence and flourishing of non-human creatures.

The third of these conditions reinforces the point that some human ambitions are to be traded off against what is necessary to enable the non-human to meet their basic needs. Some human beings will thus have to forego certain possibilities of self-development. As long as they possess that which is necessary for self-determination they will be treated in accordance with the demands of social justice, and will also be meeting the requirements of ecological justice.

Conditions (1)–(4) above thus seek to put close limits upon what the powerful demands of social justice among the most morally considerable life-forms may be allowed to justify. Human beings have morally justified calls upon natural resources. Ecologism, while fully recognising that, reminds us forcibly that we are not alone in having such morally justifiable claims, and that we cannot justify the morally simplistic view that human needs trump all other considerations. The requirements of ecological justice are also powerful.

Fortunately, some of the requirements essential to human self-determination are institutional rather than primarily material, such as the creation and maintenance of democratic systems, the possession of certain freedoms and so on. The securing of these will usually involve a minimal clash of interest between humans and non-humans.

A clear implication, however, of the approach just outlined to questions of social and ecological justice is that human flourishing is centrally a matter of securing that which is necessary for self-determination, rather than an open-ended commitment to securing human ambitions, or self-development. The important point to emphasise here is that human talents are invariably multipurpose. If an ambition which involves the use of my talent is thwarted for morally defensible reasons (I might have made a great success of slave-trading, say) then it is likely that some other outlet for that talent may be found which circumvents the moral objection. However, human flexibility is a limited commodity, more readily found in the young. Where old dogs cannot easily be taught new tricks, a case for compensation for the foregoing of a talent or skill is prima facie established, especially where matters of livelihood are concerned.

Libertarian Themes: Property and Self-ownership

The issue between Rawlsian liberals and libertarians revolves around two issues. The first is whether persons should, morally speaking, be regarded as self-owners. This concept implies that a wrong is done to them when the goods they have acquired on the basis of the capacities of which they alone are the rightful owners are removed from them and redistributed elsewhere, in the name of justice or some other moral value. The second is whether human beings should be prepared to regard property rights as absolute, so that persons' ownership rights may not be properly overridden in the name of some more fundamental moral good.

Nozick's version of libertarianism derives these claims from an interpretation of the Kantian version of the 'equal moral standing of persons' postulate we have already encountered.[15] On this view, to treat human beings as ends in themselves requires regarding them as self-owners, and the property they acquire on the basis of the capacities of which they are the self-owners as absolutely owned by them.

The first of these is intended to flesh out the idea of what is involved in autonomy or self-rule. It is supposed to preclude the Rawlsian idea that we may view the capacities with which a person is endowed as part of that person's external circumstances, to be interfered with if necessary to pursue an overarching moral aim. Nozick is happy to accept that the capacities in question are bestowed arbitrarily, morally speaking, but denies that this makes them conceptually detachable from the person whose capacities they are. Respecting persons as persons means conceiving them as specifiable, at least in part, in terms of those capacities.

The second claim, that property rights are absolute, is also viewed as what is necessary to maintain the view that persons are ends in themselves. To take away a person's property in the name of an overriding moral aim is to treat that person as a means alone. This is of course, a principle which would block redistribution of property by the state in the name of social justice.

Is there a distinctive view taken by ecologism on the issues of self-ownership and the nature of property rights? To take the former first, the following argument put forward by Kymlicka against this concept is one which ecologism would find easy to accept. Self-ownership is only a formal notion with no direct implications concerning how the world should be divided up between people. From the fact that I am the only rightful owner of my capacities nothing follows about whether the world should be held in private ownership, communal ownership or any other form. To determine the latter we have to look at other considerations, among which the most important one is to ensure that human beings each have a fair chance to achieve substantive self-determination by having sufficient control over resources to enable them to implement their projects. Hence, property rights cannot be regarded as absolute if this implies the possibility that some human beings will not have access to the resources they need to implement their projects.

This brings us to the difficult area of initial acquisition of property. Kymlicka argues that the key to the moral legitimacy of initial acquisition of property is that such acquisition should not 'deny other people's claim to equal consideration',[16] which is a version of the Lockean proviso that 'as much and as good be left for others'. He argues that the way that Locke and Nozick have got round this restriction, by allowing people to acquire access to material goods via wage labour, does not do the job required, because the point of having access to resources to acquire as property is to be able to exercise one's own autonomy, and this autonomy is diminished or lost when a person can gain resources only by submitting to the will of another in wage labour.

Thus, the question we should address when trying to decide the morally correct form of property rights to establish is not 'which property rights best protect self-ownership?', but 'which property rights best protect self-determination?' A fair distribution of property is needed for the granting of substantive self-determination to all persons, and this may well involve redistribution from the haves to the have-nots. As Kymlicka says, 'Liberal redistribution does not sacrifice self-determination for some other goal. Rather it aims at a fairer distribution of the means required for self-determination'.[17]

This kind of claim has been used to justify an argument for a universal basic income, given to people as a right of citizenship, and intended (on most versions) to cover subsistence needs. Such a solid economic foundation to the lives of human individuals is then supposed to heighten their autonomy by giving them the possibility of choosing to live on that income rather than subjecting themselves, in wage labour, to the will of others. From the point of view of ecologism, this might also have the advantage of taking some of the heat out of the forces producing rapid economic growth, among which recently heightened feelings of economic insecurity have been important. Given the point so forcefully pressed by such environmental philosophers as Leopold and Norton,[18] that the speed of economic growth is one of the most important factors threatening to overwhelm the ability of ecosystems to adapt to the human onslaught, anything which can slow down developments to a more ecologically sustainable rate will be welcome to ecologism.[19]

However, the question of whether or not a basic income scheme would heighten human autonomy, and thus one of the key elements in human flourishing, and whether it would slow down growth and thus permit ecosystems time to develop responses to human-wrought changes, are both empirical matters. Such a scheme might have no significant impact on these factors. Hence, ecologism is certainly right to give a favourable consideration to such proposals, as many Green Parties have. But since the actual impact of such schemes is a matter of speculation it is unwise for ecologism to make the idea into a cornerstone of its theorising.

To do justice to the moral considerability of the non-human, the question about when initial acquisition is morally acceptable should be framed to take account of the moral claims of other creatures to consideration. It is in this fundamental area of initial acquisition that issues of environmental justice and issues of ecological justice come together. Human beings, on this view, are morally permitted to acquire property, subject to the requirements of distributive, and environmental, justice, as long as enough and as good is left for other creatures to flourish, that is, subject to the requirements of ecological justice. However, as we have noted earlier, different species differ in terms of the degree of their moral considerability. In the case of species whose members are devoid of individuality, human appropriation which leaves the species in conditions of flourishing in viable numbers somewhere will be sufficient. But animals such as the great apes which exhibit high degrees of individuality and approach the condition of personhood need to be considered much more as individuals.

This further implies that, just as redistribution from haves to have-nots is justified within human societies in order to satisfy the requirement to treat all human beings as ends in themselves, and as capable of self-determination, so such redistribution from haves to have-nots, when it is the humans who are the haves, and the non-human which are the have-nots, is justified to treat the non-humans as morally considerable, and as having their own conditions for flourishing. Thus, ecologism justifies the redistribution of resources from humans to non-humans so as to achieve a fairer distribution of the means required for flourishing (which involves self-determination in the human case and habitat-related development in the non-human case).

This is not unfair to humans provided that they retain resources to achieve what Kymlicka calls 'effective control over one's life', to cite the formula which he employs to establish that those humans who give up property to other humans are not being treated unfairly.[20] This will require fairness of distribution of property between human beings, however, or else it may turn out that redistribution from the human to the non-human would remove from some humans the resources needed to lead a decent life with reasonable amounts of substantive autonomy. That is, since ecologism is committed to fairness in the distribution of resources between humans and non-humans, and to substantive self-determination for human beings (as a species-specific requirement for flourishing), it is also committed to redistribution between human beings.

It is, of course, possible that even when resources are fairly shared between human beings resources cannot be redistributed in favour of non-humans without taking some human beings below the acceptable level for substantive self-determination. Human interests will come first, in such circumstances, but as we have already noted in point (4) above, human beings are required to ensure that their population increase does not go so far as to create such situations unnecessarily.

Of course, when it is non-humans whose increase in population creates a shortage of the necessary resources for other creatures it starts to become possible for humans to justify measures to reduce their numbers. Non-humans are not moral agents and so cannot be expected to act in accordance with considerations of fairness. The issue of when a non-human species is no longer eligible to be given consideration because it is itself usurping resources needed for other creatures, including humans, is a difficult matter. As we noted above, however, human beings are not required to intervene to police relations between creatures which are not moral agents, except when avoidable extinctions are in prospect. When the conflict is between humans and non-humans the interest of humans plainly will carry most weight, but the interests of the non-human are to be counted too. Here we need the impartial human guardianship body, mooted in the last two chapters, to ensure that the tendency of humans to act unfairly with respect to other species is held in check.

Does ecologism, in putting forward such arguments, attack human dignity? Kymlicka argues that redistribution between human beings in order to achieve a fairer share of means for self-determination does not attack the dignity of the better-off who are made to cough up.[21] It would be such an attack, he argues, if it could independently be shown to be morally wrong to redistribute. Since, however, such redistribution can be shown to be morally required, engaging in it cannot be an attack on the dignity of the better-off – treating them as if they were slaves, to use Nozick's formulation. This argument is also available to ecologism – since it may be morally required to redistribute resources from human beings to the non-human to permit the flourishing of the latter, it cannot be an affront to human dignity to do so.

Self-ownership was designed as a concept to counter the idea that human beings have the right to enslave each other. As we have seen, and as Kymlicka shows well, it is inadequate for that purpose. For even when the concept is accepted by all human beings, this in itself is not enough to prevent other human beings from dominating you via their control of the resources which you need to live a reasonable human life, and this domination can come very close to slavery. However, although human beings may not rightfully enslave each other, it appears to be a proper question to ask whether human beings may enslave non-human creatures.

Slavery of humans by humans is blocked by the moral requirement that we respect human beings' capacity for autonomy, without which their flourishing is not possible. In the case of non-humans, however, the conditions of flourishing do not turn on such a consideration, because they do not have the capacity for self-determination. Their fate is to follow the life-pattern genetically determined for their species. Arguably the concept of 'slavery' does not properly apply to such creatures. Further, their direction and control by humans may be compatible with the flourishing of the animals in question, provided, of course, that their owners respect the conditions of that flourishing by permitting the creatures in question to live the life normal for their kind.

It is a more difficult matter to determine what is required in the way of respecting the moral considerability of creatures which have been specially bred. These may have characteristics such as docility which are then used to justify very different forms of life from those which would have been appropriate to their wild forebears. There is a whole host of issues which arises here in the wake of advances in genetic engineering. Ecologism will wish to press the issue of the moral considerability of the creatures in question and to be particularly sensitive to the possibilities of self-serving views on the part of their human owners that their property is as happy as could be in their condition of servitude – a chorus unhesitatingly sung by owners of human slaves throughout the centuries.

To return to the issue upon which libertarianism focuses, what view should ecologists take of the value of liberty? Kymlicka usefully points out that when human beings do have a legitimate aim in 'maximising' their freedom this is because certain kinds of liberty are essential to the flourishing of the kind of creatures which we are. The aim of liberty-maximisation for such creatures is, therefore, not to maximise the amount of liberty they possess, in some mysterious quantitative sense, as if you became freer the more free choices you are able to make, however trivial they may be. Rather the aim is to secure freedom in every important, or most important, respect.[22]

Ecologists take a similar view of liberty, although they, of course, broaden the issue to take account of the moral considerability of non-human creatures. Both human beings and non-humans require to be free in certain basic ways in order for their moral considerability to be properly respected. The liberties which are going to be valuable for non-humans will be a much less varied set than those needed for human flourishing. Freedom to satisfy basic physical needs for nutrition and procreation in the manner characteristic of the species, within a habitat of the sort to which the species is well or uniquely adapted, is going to be the main requirement.

It is not incompatible with this requirement that such a habitat and mode of life are provided by conscious and deliberate human action, as with the provision of a protected area of habitat within a national park. However, the dangers of eroding the freedoms of the non-human in the name of more morally pressing human requirements are always going to be large, especially in areas in which human beings are involved in micro-management of the environment. The less such management we become involved in, the greater the chances that those basic freedoms essential for non-human flourishing will be maintained. Libertarians are fond of 'keep out' signs. In the case of the liberties of the non-human they have more of a point than they often have in the human case.

Clearly, this general position does not settle what are the precise claims which non-humans have on human beings in order to secure their conditions for flourishing. This is an issue which can only be dealt with a case at a time. But it is worth noting Kymlicka's point,[23] that one of the considerations which motivates some people towards acceptance of the libertarian position is the 'slippery slope' argument which says that if you seek to equalise people's circumstances in order to ensure that the inequalities between them derive solely from the choices they make, and if you count their talents as part of their circumstances, then it is not obvious why you should not redistribute bodily parts among human beings too, where this is feasible, to equalise life-chances. Kymlicka concedes that it is a difficult issue to determine just how far equalisation and compensation may be taken in the pursuit of an endowment-insensitive theory of distribution.[24]

This concession is helpful to ecologism in one respect, namely that it shows that, however difficult it may be to determine how much redistribution or compensation may properly be exacted from human beings to allow non-humans the opportunity to flourish, it is not a difficulty peculiar to the ecologists' position. And in one respect ecologism has a slightly easier time in making some decisions. For example, it is clear that ecologism is not committed to arguing that human beings should participate in a 'survival lottery' with other life-forms, so as to make some human beings available to the smallpox virus, say, as part of an exercise in equalising life-chances. For, as we noted earlier, moral considerability is not the same as moral equality, and some less morally considerable life-forms may be properly sacrificed, or even exterminated, in order to preserve the conditions of flourishing of life-forms with greater degrees of moral considerability.

Notes

1. Will Kymlicka, *Contemporary Political Philosophy* (1990).
2. Kymlicka 1990, p. 5.
3. Kymlicka 1990, p. 24.
4. We have already touched on a version of this issue in discussing in Chapter 5 the claim of Lynch and Wells that human beings as such are in a special relationship with other human beings.
5. Kymlicka 1990, p. 42.
6. Kymlicka 1990, p. 11.
7. John Rawls, *A Theory of Justice* (1972).
8. In what follows the complications created by the point made at the start of Chapter 5, that personhood admits of degrees, are ignored.
9. Rawls 1972, p. 3.
10. Kymlicka 1990, p. 161.
11. Ronald Dworkin, 'What is equality?' (1981). Cited by Kymlicka, on p. 75.
12. Rawls 1972, p. 73.
13. Rawls 1972, p. 4.
14. Ted Benton, *Natural Relations* (1993).
15. Robert Nozick, *Anarchy, State and Utopia* (1974).
16. Kymlicka 1990, p. 109.
17. Kymlicka 1990, p. 122.
18. See Bryan Norton, *Toward Unity among Environmentalists* (1991) on this important theme in which he echoes the ideas of Aldo Leopold on the need to adapt human life to the time-scale of biological processes.
19. P. Van Parijs (in Van Parijs (ed.), *Arguing for Basic Income* (1992), p. 27) suggests that the popularity of basic income schemes among Green Parties is to do with the fact that such parties contain disproportionally large numbers of people with a preference structure which privileges leisure over material consumption. This may well be so, but this claim ignores the question of why that preference structure might be deemed more reasonable, from the point of view of environmental theory.

    The volume edited by Van Parijs admirably demonstrates the variety of arguments which may be marshalled to support the idea of a basic income. There is not scope in this book to explore these in detail. But one point it may be pertinent to make here is that the idea of an 'income' is not one whose usefulness is restricted to the human case.

Taking an income as 'a stream of resources which permit the maintenance of life-processes', every living thing needs an income. If there is a case to be made for a basic income for human beings there is equally a case to be made for a basic income for non-humans.

Whereas in the human case this requires that societies organise their economic and distributive processes in certain specific ways, in the case of the non-human it is a matter of human beings engaging in certain kinds of forbearance, so as to permit non-humans access to their own income streams. Certain other themes central to the discussion of basic income for human beings also drop out of consideration when we turn to the non-human, such as the issues of laziness, fecklessness and conspicuous consumption.

20. Kymlicka 1990, p. 120.
21. Kymlicka 1990, p. 123.
22. Kymlicka 1990, p. 151.
23. Kymlicka 1990, p. 155.
24. Kymlicka 1990, p. 155.

9

Ecologism and Contemporary Political Philosophy: Marxism, Communitarianism and Feminism

Marxist Themes: Exploitation and Alienation

THE FIRST ISSUE TO which a discussion of Marxism directs our attention is that of the 'circumstances of justice'. This is the issue of when it becomes a relevant matter to determine rules whereby benefits and burdens are to be distributed within human societies. As Kymlicka points out, the traditional view on this matter is that we need rules of justice only because human beings have conflicting goals, and we are faced with a situation of limited material resources and thus have to determine how it is decided what is done with these resources – who is to benefit from them and how? The implication is that if we could all agree on goals which we jointly sought to achieve, and if we could attain a situation of superabundance, so that resource issues did not have to be addressed, then issues of justice would not arise.

Marxists traditionally did believe that these circumstances of justice could be eliminated. Most now realise that this is not possible, and thus some hard thinking about rules of justice is required. What does ecologism maintain on this matter?

As we noted in Chapter 5, Barry has demonstrated that issues of justice arise even outside the circumstances of justice, so that Marxists, along with Hume and Rawls, are incorrect in supposing that the circumstances are a necessary condition of the question of justice arising. It may still be argued that the circumstances are a sufficient condition for such issues to arise. As far as ecologism is concerned, it clearly has no difficulty in accepting the claim that human beings are inevitably faced with a situation of scarcity of resources, notwithstanding the claims of defenders of neo-classical economics that the most important resource, human ingenuity, is limitless.

However, even if the latter claim could be conclusively established, and thus it could be demonstrated that no practical resource limit exists for human beings, the second circumstance of justice would still provide ecologism with its case for ecological justice. Ecologism clearly must emphasise the impossibility of achieving a complete harmony of goals among morally significant beings. Human beings are unlikely to agree among themselves on such a harmony, for human beings are a highly individualised species, which means, for ecologism, that such a harmony could at best be a matter of chance and inevitably transient.

However, when we bring into the picture the goals and interests of the non-human then there is no agreement possible even in principle, for humans and non-humans cannot reach an agreement on anything. There is clearly also no natural harmony of goals between human and non-human. Thus, for ecologism, these 'circumstances of justice' are written into the nature of life, and thus human beings, as the only moral agents in this situation, cannot avoid the task of seeking to reconcile justly the inherently conflicting purposes of human beings among themselves and the conflicting purposes of humans and non-humans. The existence of conflicting purposes between non-humans is not an issue of justice, for no immoral action, and so no injustice, is possible between beings which are not moral agents.

Marxists, of course, anticipated a future form of society in which, once the circumstances of justice had been overcome, the place of justice would be taken by concern for the interests of others based on altruism, or a spirit of fraternity. Corresponding to this conception is the phenomenon of love and fellow-feeling for the non-human world which we have already introduced on several occasions, namely biophilia. However, this is not something which ecologism would wish to see replace considerations of justice. For, as Kymlicka points out *vis-à-vis* the inter-human case, we may need to ascertain what justice requires us to do in cases where we are motivated by love, but face conflicting demands created by that very love.[1]

This applies also to the relation between humans and non-humans. Indeed, a clearly articulated and intellectually defended conception of justice is arguably even more important in the human/non-human case, given that our mutual affection as human beings may lead us to ignore or denigrate the interests of the non-human more readily than we do, so lamentably often, with members of our own species.

The main Marxist charges against capitalism, of course, have focused on two rather different concepts from that of justice, namely exploitation and alienation. The concept of exploitation has been subject to much detailed discussion in recent years, and it has emerged in the course of that discussion that the Marxists' preferred remedy for exploitation, namely socialising the means of production, is neither a sufficient nor a necessary condition for the ending of exploitation, when this is understood as the unfair use of one's power to gain an advantage over others.[2] Certainly, ecologism can accept the specification of non-exploitative relationships in the economic sphere given by Kymlicka, namely that these obtain when people are able to make their own decisions concerning the conduct of their lives because they are able to obtain the material means to implement those decisions.[3]

Ecologism needs to be concerned about inter-human exploitation as part of its general concern with the conditions needed for the flourishing of all creatures. Given the human capacity for self-determination as part of the conditions for human flourishing, the creation of a system of relationships within the human economic sphere which are not exploitative is a matter of central concern to ecologism.

However, prima facie, such access to resources may be achieved in various ways, including the creation of a property-owning democracy within capitalism, market socialism, or social ownership. Ecologism needs to be pragmatic in this area, which is littered with rigid dogmas. Arguably, ecologism should take the view that the need continuously to produce ways of maintaining living standards so as to enable human beings to live a decent life while simultaneously living 'lightly' on the Earth requires continuous technological innovation of the kind of which only the market system has in practice been the provider. We return to a fuller discussion of these matters of political economy in Part Four.

How do ecologists view exploitation in the human/non-human context? On the definition of 'exploitation' given so far, human beings would be said to exploit non-humans in the morally objectionable sense if they were to use their power over non-humans unfairly to gain an advantage over them. Such unfairness in turn, on the analogy with the purely human case, would involve a morally indefensible unequal access to the means of production (or means of life in the case of species which do not 'produce' in any meaningful sense) which favoured human beings. Non-exploitative relations in the human case would involve, as we noted, the ability of each human being to have access to the means of production in such a way as to be able to make decisions concerning work, leisure and risk which suited each person's goals.

None of this transfers easily to the non-human case. Non-humans do not have the capacity to make any decisions concerning the balance of work, leisure and risk within their individual lives, although many higher animals do have complex lives in which it is possible for human observers to discriminate periods of effort, including tool use, from periods of leisure, and it is not totally impossible to determine that such an animal has made a calculation of risk, as when it moves to drink from a waterhole in clearly nervous apprehension of the possibility of predation. But arguably no non-human animals have the capacity to take a longish-term view of their lives as such and weigh these factors against each other.

Even the idea of making decisions is hard to apply to many non-human life-forms. Many creatures simply live as best they can the lives ordained for them by their genetic constitution, with very little adaptability built into their repertoire. However, it is possible to view this, as Popper has suggested, in terms of the embodiment within the genetic constitution of the animal of a theory about how best to live. Such theories may be falsified, in which case the species will perish. The advantage which human beings, as persons, possess over other animals on this planet is that we can propound and test theories without risking extinction if they go wrong.[4] However, this advantage, from the ecologist's point of view, does not morally justify ignoring such 'embodied' decisions about a creature's life-plans. A non-exploitative relation between humans and non-humans, therefore, will be one in which the 'genetic decision' of non-humans concerning their life-goals (mainly, of course, reproduction) are treated as being as worthy of consideration as the conscious decisions of human beings *vis-à-vis* their chosen goals. This implies, as before, a single principle, of respect for life-choices, with different implications for different species.

Turning now to the issue of alienation, the first point to make is that this is a notion which has perfectionist implications. In other words, if a claim is made that human beings are in a state of alienation, then this implies that there is a disalienated state in which human beings are able to experience the characteristic excellences of their species. As Kymlicka explains, in the case of Marxist theory the characteristic excellence was held to be the exercise of the capacity for freely creative co-operative production.[5] Marxists have often been accused here of elevating their own preferences to a universal truth. Kymlicka's reaction is an example of this criticism, for he says 'there is no reason ... to exclude or stigmatise those who prefer the passive pleasures of consumption over the active pleasures of production'.[6] Freely creative production is one, but only one, way of being distinctively and rewardingly human.

Is ecologism perfectionist about human life and/or non-humans? As already noted, the perspective of human ecology recognises that human beings are highly individualised and that each human being has the capacity for autonomy, so that no single form of life is likely to conduce to the flourishing of all individuals of the species. With respect to non-humans, however, their lack of individuation and autonomy makes it much easier to specify a mode of existence within which the perfection of the creature's life is attainable. In so far as the creature is able to lead the life for which it has been adapted, within the appropriate habitat, its attainment of maturity and its carrying out of the functions normal for a specimen of that type amount to a perfection of its existence.

The failure to attain these conditions leads to a sickening of the creature, and in the case of higher, more complex animals, the examples of specimens kept away from their habitats in zoos suggests that it is not inappropriate in certain situations to speak of their existence as alienated. This may even be true of specimens kept as pets which are well looked after and in good physical shape. After all, it is part of the concept of alienation that creatures suffering from it, even human beings, may not be aware of their condition, and may, from their own viewpoint, feel perfectly content. However, to the extent that an animal species approaches the condition of personhood, showing a complex repertoire of behaviour and great adaptability, it becomes possible to argue that there is a variety of modes of life, including ones lived in highly interactive relations with sympathetic human beings, which are fulfilling for members of that species. Perfectionism begins to recede as a theoretical option, or at least becomes more complex, in such cases.

There is a fallacy to be avoided in the discussion of the

alienation/perfectionism nexus, and that is the supposition that a good or perfect life for a given species, human beings included, is one which develops the capacities which are unique to that species. This is the fallacy which many, including Kymlicka, detect as being present in the Marxist discussion.[7] What differentiates a species from other species need not have any part to play in the specification of a, or the, way of life excellent for that species, where it makes sense to speak in such terms. It is part of the general spirit of ecologism, which views human beings and non-human species as closely connected, not divided by an unbridgeable chasm, that it is ready to accept that what is a condition of fulfilment in at least many human lives may be something human beings have in common with all, or many, other creatures. Interaction with a rich natural environment may well be such a common feature.

Marxists have always prided themselves on having a clear and, at least on the face of it, plausible account of the agency by which social change in the direction of a non-exploitative and disalienated society may be achieved. The role of the proletariat in effecting the revolutionary transformation of the historically necessary stage of capitalism into the socialist-communist era appeared to rest on the hard, simple, self-interested calculations which Marxists set so eloquently before it.

It is not easy to identify any plausible groups of that kind in ecologism's prognostications about how the changes which it seeks will come about. Of course, if the biosphere is facing as serious a crisis as many political ecologists claim, then all humanity has a direct, practical interest in doing what is necessary to overcome that crisis. But, as we have frequently noted, the diagnosis of crisis rests upon scientific analysis which, in the way of all such analysis, is open to alternative interpretations, a fact which Marxists failed to see applied to their own theory, even granting its scientific status. Such alternative interpretations serve to permit a reasoned scepticism about the ecologists' prognoses of a kind which, even if they are true, may well hinder their acceptance until it is too late.

However, let us recall that ecologism, unlike at least one influential version of Marxism, often put forward by Marx himself, is a self-consciously moral position, resting on the basic value-claim of the moral considerability of the non-human. Thus, while many may be brought to produce the kinds of change which ecologists desire for reasons of enlightened self-interest, from the point of view of ecologism only those who do so in pursuance of the value-postulate will be doing so for the right reason.

Which groups of humans, if any, are likely to accept the value-postulate? Some may accept it as the result of their commitment to a religious position. But ecologism is only going to grant that the value postulate is securely anchored if it rests upon an informed grasp of the biological realities within which we live, as provided by scientific ecologism and allied disciplines. Thus, acceptance of the right principle, for the right reasons, requires that people have sufficient intellectual sophistication and knowledge to grasp the biological theories which underpin ecologism. They will also need a certain, perhaps rare, kind of moral imagination to be open to the claim of the moral considerability of the non-human. Their sense of the wonder of life will also need to be intact and in a reasonably heightened condition. And then they need the energy and commitment to play their part in the business of politicking, dissemination of ideas and formulation of practical policy.

It is clear that there is no distinct sub-group of humanity among which these characteristics may reliably be found, notwithstanding the theorising about the role of the new (middle) class in advanced industrial countries, the participants in New Social Movements.[8] Such people do exist, and are important. But it is clear that the political aims of ecologism will be achieved only by a pragmatic working-together of various groups – ecological ideologues and interest-groups, environmentally sensitive businesses, political parties and leaders, third world peasant groups and so on, all motivated by different considerations.

This looks to be too much of a rag-bag to convince those looking for a successor to the proletariat as the agents of revolutionary change. However, given that the proletariat has so comprehensively failed to carry out its historical mission in the capitalist heartlands, this eclectic, open-ended, loose coalition of groups and interests may actually be a more promising basis for securing effective change of the kind sought by ecologism. The variety of interests and motives involved may maintain the impetus for environmental change should any one of the strands lose, perhaps temporarily, its effectiveness. The danger for ecologism, as we noted in Chapter 4, is that support for the policies advocated by ecologism, if it rests on reasons other than those which ecologism cites, may lead on in directions undesired by ecologism. Supporters of the ideology need to be constantly on the *qui vive* to ensure that the ideology's aims do not become side-tracked in this way.

A final point to consider before leaving Marxism is the claim made by Kymlicka and others that the positive value attached by Marxism to disalienated labour is not the only view which may reasonably be held concerning the role of productive work in human life.[9] Another reasonable view, for example, is that the real point of labour is the efficient production of what is necessary to meet vital needs. From this standpoint, alienation in the workplace may be a price worth paying if such production enables more human beings to meet their vital material needs than would otherwise be possible.

This is a general position which ecologism supports, for ecologism itself makes the point that one of the main purposes of human labour must now be to produce the means to achieve human material well-being without destroying the conditions for the well-being of other species. The organisation of work among humans should therefore aim not simply to reduce alienation and exploitation where it can and meet human material needs, but also bear in mind what is needed to do justice to the moral considerability of the non-human. Thus, human labour does indeed have many purposes, of which the provision of disalienated forms of productive activity may, at this stage in our planet's history, be one of the less important ones.

Communitarianism: Autonomy and Tradition

The challenge which communitarianism poses has been directed mainly to the tenets of liberalism, particularly to the latter's understanding of what the concept of individual autonomy involves and the conditions for its attainment and exercise.[10]

The main thrust of the communitarian position can be summarised in the following four propositions:

(1) individuals cannot achieve meaningful autonomy by themselves, but derive the values and purposes definitive of their projects from the social setting in which they come to maturity;
(2) there is no transhistorical source of such values, rather they all come from concrete social contexts;
(3) each social context, embodying traditions and values, customs and institutions, in effect endorses a specific vision of the good life;
(4) therefore there is no neutral position for a government to

> occupy *vis-à-vis* the fundamental values of the society – its basic task must be the maintenance and furtherance of the specific vision of the good life inherent in the society it governs.

Kymlicka sets about providing a liberal counter to these claims. He first of all denies the claim, implicit in (4), that liberalism is committed to the position that one view of the best life for human beings is as good as any other.[11] In principle liberals concede that a government could work out what is for the good of everyone and impose it upon the citizenry. However, what liberalism objects to with respect to this possibility is the idea that persons' lives can be made to go better by the imposition of values they do not themselves endorse. For the liberal, then, a necessary, though not sufficient, condition of one's life going well is that one is living it 'from the inside', in accordance with the values one really endorses.

Some forms of paternalism are compatible with this. The government in a liberal society may properly compel people to do things which those people endorse 'from the inside' in their reflective moments, but tend to ignore in practice, through inadvertence or backsliding. It may also, as a short-term expedient, encourage or require people to engage in activities in order to introduce them to ways of life which are valuable in the hope or expectation that those people will come to accept that value.

But in general, the liberal case for individuals to have freedoms which enable them to try out different forms of life and to hear criticisms and suggestions from their fellow-citizens about what is good in life derive, not from the view that there is no truth of the matter to be found, but rather from fallibilism about the claims which people make. The truth, in the matter of the good life for human beings as in the matter of the truth of scientific claims, requires a regime of free enquiry coupled with searching scrutiny – both requiring a government which permits its citizens the maximum freedom of all, compatible with equal freedom for each. This is an argument which we have seen ecologism endorse as part of the case for democracy.

Thus, Kymlicka argues, to affirm the neutrality of the state is not to reject the idea of a common good, but to provide an interpretation of it.[12] For liberals, the common good emerges as the result of the working of political and economic processes which combine individual preferences into a social choice function. By contrast, on the communitarian view of society, the common good is conceived of as a substantive conception which is definitive of the community – point (3) above. Communitarians within a given society, therefore, assess proposals for new ways of life in terms of how well they fit in with the existing values and associated practices of the society.

On this dispute, ecologism finds reasons for criticising both positions. First, it will reject the communitarian view that traditional practices of human societies must be granted a sacrosanct status, for many of the traditional practices and values of human societies across the world have played a large part in creating our present environmental predicament, specifically by failing to recognise, or to act fully on the recognition of, the moral considerability of the non-human. Secondly, it will want to restrain individual and group preferences within liberal society, and put restrictions on the production of the social choice function by economic and political processes, on the basis of what is needed to recognise the moral considerability of the non-human.

Ecologism clearly does need to persuade people to accept its key value-claim and internalise it, so that it guides their view of the good life. If this is not achieved, then any advances it may obtain by the imposition of its value-postulate and implied prescriptions would be inherently short-lived. Just as important, since autonomy is, from ecologism's point of view, a condition for the flourishing of the human species, and since ecologism is committed to securing the conditions of flourishing for all beings as far as possible, it will be committed to securing the conditions for human autonomy, including the internal acceptance of key values.

This, of course, means that political ecologists will have to accept that some, and perhaps many, human beings will exercise that autonomy in the pursuit of values, such as gross consumerism, which threaten to lead to the extermination of other life-forms on a large scale. There is, however, no alternative to the business of seeking to change values by persuasion, not imposition, even though there is no guarantee of success. The freedoms characteristic of liberal society are essential to this, and so ecologism will staunchly defend those values as a key part of its own political theory. The two kinds of paternalism noted above will, however, be available to ecologically committed, democratically accountable, governments and thus enable some defensible forms of coercion to be applied in support of ecologism's aims.

Further, ecologism will support the liberal view expressed by Kymlicka, namely that 'no particular practice has an authority that is beyond individual judgement and possible rejection'.[13] Ecologism, as with any radical theory, must endorse this as a key element in the critical rejection of many existing social, economic and political practices. As Kymlicka concedes, we cannot meaningfully reject all our commitments at once, for we need to take something as given in order to be in a position to reject, rationally, something else. However, no commitment or project is inherently unrejectable. Over time we could end up with a completely different set of commitments from the ones with which we began. This, Kymlicka argues, is the sense in which the self is 'prior' to its commitments – not that the self could simultaneously stand outside all its commitments, but that over time a given self could transform all its commitments and still be said, meaningfully, to be numerically the same self throughout.[14]

One problem which besets liberalism as a result of its commitment to the neutrality of the state also needs to be addressed by ecologism. This is the problem that, as Raz has argued, pluralism may be self-defeating.[15] This is because a state in which the government maintains neutrality with respect to competing conceptions of the good cannot ensure that a diverse culture capable of offering its citizens a wide variety of options will remain in existence. Kymlicka concedes that this does restrict the neutrality of even the most liberal state. In one of his few forays into the discussion of environmental issues, he gives the following example: The state may need actively to protect wilderness areas so that future generations may have the choice of enjoying them. Left purely to individual choices, people may so act as to ruin or use up the resources needed to sustain a practice in the future, thereby reducing future choices. However, he argues that the state's preservation of the fullest range of options is not the same as the state's actively promoting any of them.[16]

Ecologism's view on this matter must derive from its commitment to the moral considerability of the non-human. Let us recollect that liberalism has a commitment to the equal moral standing of all persons, and that the neutrality of the state is to be regarded as the best means to securing this equality. This means in turn that the liberal state is committed to limiting the activities of groups which use the freedom they obtain on the basis of the basic moral principle to attack the equal moral standing of others.

Ecologism is committed to the moral considerability of the non-human, as well as to the equal moral standing of all persons. Thus, it is committed to restricting the activities of human beings who use the liberty granted to them by the 'moral equality of humans' principle to attack the moral considerability of the non-human. So the issue of pluralism for ecologism is not simply that of how to keep options open in order for human beings to have choices available to them, but how to maintain life-chances for non-human species as part of the recognition of their moral considerability. Thus there will be rather more limitation on the neutrality of the state under ecologism than is envisaged under liberalism.

However, there is one area in which the neutrality of the state under both liberalism and ecologism cannot be expected, namely with respect to the basic moral postulate which each supports and which forms the basis for the liberal or ecological state's own legitimacy. Both political ideologies have a commitment to pluralism, for reasons of fallibilism and the preservation of autonomy, which involves allowing challenges to be made even to their own fundamental moral postulate. But, committed as it is to the truth of that postulate, each theory has to restrict the amount of influence the detractors from its own moral position are to be allowed to have.

Thus, liberals may be required by their own postulate to permit the expression of fascist viewpoints within liberal society to a certain point. But beyond that point, suppression of those views starts to be a requirement on the basis of the same postulate, on pain of allowing the liberal society to disappear. This is the implication of the view, brought out by Kymlicka, that liberalism does have a view of the good life. At some point liberals have to be prepared to fight for that view.

Similarly, ecologism can permit the expression of anti-ecological viewpoints, for the fallibilism and autonomy reasons already noted. But only to a certain point, for reasons parallel to those given in the case of liberalism. These points are logical ones. They emerge from the idea of what it is to be committed to a substantive moral position. In practice, of course, there may be no resistance within an initially ecological or liberal society to the spread of human chauvinist or fascist ideas, either through lack of real conviction, timidity or errors of judgement on the part of political ecologists or liberals, or because of the genuine conversion of political ecologists and liberals to their opponents' position. But the logic of the argument remains, and with it the uncomfortable issue of trying to decide in practice when to permit expression of hostile ideas and when to suppress them, carrying with it as it does the inevitable, but confused, charge of hypocrisy.

Kymlicka argues for a distinction between two kinds of perfectionism.[17] Liberals, he says, support 'social perfectionism' in which the good life emerges as the result of human beings' pursuit of their own visions of the good life within civil society. Communitarians support 'state perfectionism', in which the state is supposed to promote a substantive vision of the good life. In terms of this distinction, ecologism lies somewhere between the two extremes. It tends towards social perfectionism with respect to human beings. But with respect to the non-human it has, *faute de mieux*, to take a state perfectionist view. This is because non-humans cannot themselves articulate a view of the good life. The ecological state, through its guardianship institutions, has to articulate it on their behalf and ensure that it is fairly taken account of in human deliberations.

However, the tasks of an ecological government may be made easier by ecologism's acceptance of the human capacity for self-rule, which underlies its case for democracy, discussed in Chapter 7. This capacity implies that the business of deciding on values and options is one which may be conducted on the basis of collective activity in civil society, rather than on the basis of political activity, organised through the state. A citizenry which is committed to ecologism is one in which the business of attending to the implications of the basic ecological value postulate may be widely shared.

Ecologism shares the view which Kymlicka argues is held by liberalism, as against the criticism by communitarians of liberalism expressed in (1) at the start of this section, that human beings, being a species of social animal, 'naturally form and join social relations and forums in which they come to understand and pursue the good'.[18] They do not, therefore, need the state to make them do that. The role of the state is rather to maintain the moral limits on the social deliberations of human beings. Thus all political theories are, to revert to Kymlicka's distinction, 'state perfectionist' with respect to their fundamental moral postulates.

Kymlicka correctly notes that the historical examples cited by communitarians, in support of their view that shared values are necessary for a political system to be accorded legitimacy by the citizenry, are of societies in which an artificial homogeneity of values was achieved by the exclusion of groups such as women, slaves and foreigners whose values would have conflicted with those of the politically privileged groups.[19] On this basis, Kymlicka criticises communitarianism as being remarkably ignorant of the history which is supposed to form the basis of the theory's distinctive position.[20] Ecologism will wish to broaden this criticism, with which it can readily agree, to liberal societies. For the latter have also systematically excluded the interests of creatures outside the privileged, human, groups.

Kymlicka, in addressing another communitarian criticism, this time of what is supposed to be the inherently fragile nature of liberal societies, also argues that it is not necessary for a society's members to share a substantive view of the good in order for them to be willing to make sacrifices in order to support the have-nots in their society. All that is needed is a shared commitment to justice, a procedural and formal conception, to bind a society's members together. Ecologism will find this a congenial position, with the caveat that the just procedures in question will require to be extended to encompass ecological justice.

Whether a shared sense of justice is enough to bind human societies together is still a moot point. But liberal societies, for all that there is a pluralism of values within them, do not seem less stable than societies which, on the face of it, share more substantive commitments, such as religiously based ones. In any case, the idea of shared values is itself one which rapidly becomes complex under scrutiny. For example, the sharing of values by Christians has not stopped them disagreeing violently over interpretations of those values and the prioritisation of them in situations where they cannot all be realised simultaneously.

As Raz has argued,[21] people can and do accept that their state supports values with which they are themselves not in agreement, as long as this support emerges from an agreed procedure for arriving at a public ranking of the values of different ways of life. If this argument is correct, then it strengthens the view that a state committed to ecological value-positions need not seek to gain the acceptance of those values by all its citizens on pain of forfeiting its legitimacy. Ecological societies, thus, do not have to be viewed as inevitably committed to a totalitarian posture.

Feminist Themes: Public versus Private; Caring versus Justice

•

The first issue to be addressed in this area of political theorising is the dispute between 'difference' and 'domination' theories of female liberation. The former sees the political problem facing women as being that of overcoming arbitrary prejudices against them in education, employment and social life in general, so that women and men may compete fairly with each other for the social goods which society makes available. The 'dominance' view is, by contrast, that the social, economic and political world is dominated by the male sex in such a way that the very structures of work and politics have been created with the needs and characteristics of only one sex in mind.

Thus, on this view, even when arbitrary prejudices are swept away, women remain at a severe disadvantage compared with men in their careers and life chances. Kymlicka argues that the 'dominance' approach is one with which the underlying egalitarian commitment of liberalism to treat all persons as moral equals is entirely compatible. This is because that principle 'asserts that women's interests and experience should be equally important in shaping social life'.[22]

Ecologism is particularly sensitive to such 'dominance' approaches to the understanding of how the requirements of morality may be subverted. This is because it emphasises the ways in which the moral considerability of the non-human has been systematically ignored as the result of human domination of the planet. Some ecofeminists have challenged this parallel, arguing that it is male domination which underlies both the subordination of females and the domination of the non-human. Thus, on this view, it is not the human species as such which is the culprit in the environmental sphere, but the human male, whose dominance has been maintained in part on the basis of the elaboration of various dualisms, such as that between reason and emotion, in terms of which the claim for the moral superiority of males over females and of the (male) human over the non-human has been promulgated. Such feminists bolster their position by pointing to the way in which nature is often personified as a female to be subdued to the wishes of her natural master, man. This analysis implies in turn that the overcoming of such dualisms and the emphasising of female 'caring' and nurturing characteristics are needed for both the liberation of females in human society and the liberation of the non-human from the depredations of male domination.[23]

This version of the domination argument has the advantage that it seeks to establish a logical connection between the two kinds of domination. Failing such a connection it appears that there is no reason to suppose that a commitment to ending the domination of females by males entails a commitment to the ending of the domination of the non-human by the human. The two issues would be logically distinct.

The question of whether or not there is such a logical connection, mediated via pernicious dualisms, is still a matter for debate. However, even if the logical connection cannot be established, it is still possible for ecologism to link the two kinds of domination, on the basis of its approach to morality. This approach, as we have frequently noted, sees the discussion of intra-human moral issues as a subsection of the wider sphere marked out by the notion of moral considerability. The domination of human society by one sex, to the detriment of the conditions of flourishing of the other sex, and the domination by humans of the biosphere, to the detriment of the conditions of flourishing of the non-human, are part of the same phenomenon of moral chauvinism. There may not be a logical connection here, in the sense that the same (dualistic) theory is involved in the justification of both kinds of chauvinism, but the political ecologists' perspective at least sees a connection in terms of a moral fault common to both positions: they are morally connected, even if not logically connected.

The difference approach is obviously one which will not apply to the morality of human/non-human relations. The human and the non-human do not form members of the same society, competing for life-chances in the same forums. Hence there is no sense to the idea of removing arbitrary prejudices in that competition. But it may be asked whether the dominance approach transfers any more successfully to the human/non-human sphere. The dominance approach as applied to the issue of the relations between the sexes requires the concept of the structuring of social arrangements so as to favour one sex over the other. What is it that is supposed to be unfairly structured when it is the human/non-human we are considering?

Human beings and non-humans do not inhabit the same social space, but they do inhabit the same moral space. Only humans can become aware of this, however. This clearly provides them with the opportunity to structure that space in a way which systematically favours them at the expense of the non-human. Initially this is a matter of simply claiming that only human beings occupy moral space, which is akin to the 'privatisation' of females in human society. The latter involves claiming that the public sphere of human societies is properly occupiable only by males. Once it is acknowledged by human beings that non-human life-forms are also occupiers of moral space, then it remains possible for human beings, who are the only beings able consciously to explore and operate within that space, to structure it unfairly in their own interests.

This is a matter of denigrating the moral considerability of the non-human, and of deeming human needs and interests to be of almost infinite weight as compared with those, reluctantly admitted to exist, of non-humans. But this denigration and downplaying of the non-human is, crucially, not done to keep the non-human in its place directly, in the way that humans, males and females, are said to have propagated myths about female inferiority in order to keep women in their place. For, obviously, non-humans cannot be addressed in this manner. The anti-non-human rhetoric is, of course, directed at those human beings who do wish to see moral space structured more fairly.

A key ingredient of the 'dominance' approach to issues of sexual equality is to challenge the traditional view of the family involved in male chauvinism. This sees the family as a purely 'private' sphere, properly beyond the reach of governmental intervention, within which the female caring and nurturing role has its proper scope, albeit in such a way that the role of the male as the head of the family, representing it to the outside world in the public space, is also given its due place.[24]

Kymlicka correctly points out that liberalism's commitment to a distinction between the state and civil society was intended to protect society from state interference as far as possible, not to delimit an area of the purely 'private' from a public space. The sphere of civil society is itself a public sphere, within which there is a domestic sphere, the family, and the purely individual or private realm, within which people may escape from the social entirely, for a part of their lives. This sphere of the purely private may correctly be demarcated within the family, for just as the state and the wider society may improperly interfere in the domestic matters of the family and the lives of individuals, so families may contain forms of domination of their members, who thus require protection by the marking out of a private space within families themselves.[25]

Thus, the important issue is how such lines between the state, society, family and individual spheres are to be drawn. This is not a matter of drawing a single line between the public and the private, and simultaneously labelling the public sphere as one of a 'higher calling' within which only superior beings – males – may be allowed to operate. The liberal view of these matters has been to reject the ancient view of politics and the state as a higher calling, and instead to regard it as a necessary evil, required to maintain a complex social structure within which the individual and the social are allowed to operate within their appropriate sphere. This, Kymlicka argues, is the dominant view of the state now, and thus the only live question is what kind of structure is to be held in place by the political. Women cannot properly be excluded from the public and political and confined to the private and domestic, not simply because that rests on a false view of the abilities and appropriate roles of the sexes, but because the view of the state as a higher sphere upon which it rests is no longer tenable.[26]

Ecologism certainly views the role of the state as the important one of regulating civil society, the domestic and the private in a way which supports the moral equality of all persons. But it also sees it as having the vital task of regulating these spheres in order to support the moral considerability of all living creatures. This involves a kind of higher calling, for it amounts to the task of defending the moral claims of creatures which are not capable of articulating those claims themselves. That is, the state is not simply entrusted with a 'ring-holding' task of setting up and maintaining the ground roles governing human interactions.

Some might want to argue that the distinctive 'caring' approach of women suits them more easily to this role of 'guardian of the moral considerability of the non-human'. However, this is an empirical claim of which ecologism will tend to be sceptical. It will be recollected from the discussion in Chapter 3 that the evolutionary basis for biophilia derives from the immense time humanity spent in hunter-gatherer societies. It is clear from an examination of the roles of the sexes in those societies that both men and women needed to develop similar attitudes towards the non-human.

Political ecologists tend also to be sceptical of the view espoused by some feminists that males and females are different kinds of moral being,[27] the former supposedly being devotees of an ethic of justice, characterised by a commitment to rationality and impartiality, while the latter are supposedly committed to an ethic of caring, characterised by contextual, particularistic, narrative-based modes of moral thought. Whether or not this claim of moral psychology is true, Kymlicka has in any case some telling points to make against the view that there is a viable distinction here. For example, he points out that attending to a complex situation is a matter of selecting the features to attend to, and to do this one needs some principled basis for making the selection. Then there is the difficulty that attending to existing, concrete relationships is inherently exclusive – moral claims of strangers do not get a look-in. The attempt to cope with this by arguing for the web of relations to be spread as far as possible is hard to justify without referring to an underlying universal principle of equal moral worth.[28]

For ecologists, the problem with the idea of an 'ethic of caring' in application to the moral considerability of the non-human is that very few non-humans enter into any relationships with humans. It is also easier for humans to care for some non-humans than for others, mainly those which can offer some form of reciprocity. Then there is the crucial point that we can only properly exhibit care for a being if we are aware of what its interests are, of what counts as harming or benefiting it. For the vast majority of non-human creatures an awareness of their interests is highly dependent on a grasp of theory rather than a matter of what emerges via concrete, context-laden interaction with them. In the modern, increasingly urbanised world, such chances for concrete relationships with the non-human are in any case becoming harder to come by.

One distinction between the ethic of justice and the ethic of care which Kymlicka usefully brings out is that the former only commits us to trying to do something about the relief of distress or suffering if it has come about as a result of injustice. It regards some suffering and distress as the responsibility of moral agents themselves. It is, he argues, unfair to impose on others the requirement that they attend to your hurt and distress (except in certain extreme, 'Good Samaritan' cases) when that has arisen from your own fault.[29]

By contrast, the ethic of caring makes no such distinction – the distress of others makes a prima facie call on us, whatever its cause. This threatens to suck us into a life devoted to nothing but the care for others, since there are so many causes of hurt. So we need some principled way of determining which kinds of hurt do make a call on us, both to preserve some autonomous existence at all and to enable each of us to know when we may properly ask for and expect help and when we may not. This information is needed in order to make a proper assessment of risks involved in various courses of action.[30]

Kymlicka concedes that this position presupposes that we are dealing with able-bodied, mentally competent adults interacting in public life. He comments that 'Children cannot reciprocate care equally, they require a degree of selfishness and attention that is specific to them'.[31] This shows how an ethic of care can be relevant to our interaction with beings not capable of (fully) reciprocating our care. However, it also suggests that human beings can only cope with this non-reciprocity (and even then, not all do so successfully) on the basis of the considerable natural affection they are inclined to feel towards their own offspring.

It is clear that this is not going to be forthcoming with respect to the vast majority of non-humans. This gives us another reason for arguing for a principled basis for the care of non-humans. Biophilia may get us some way in situations of non-reciprocity, but given the size of the demands which the recognition of the moral considerability of the non-human is likely to impose on us, the role of universalisable moral thinking central to the ethic of justice needs to be emphasised.

Kymlicka ends by conceding to the devotees of the ethic of care that those who have argued for an ethic of justice have done so in such a way as to suggest that adult human beings do not have dependants whose care has to be organised. Thus they have neglected the issues of how care is to be distributed, which is itself a matter of justice, assuming too readily, as feminists charge, that women will automatically do the caring within the domestic environment deemed to be beyond the scope of issues of justice. In doing so they have neglected to consider how individual autonomy of adults may properly be limited by the need to engage in caring for dependants. In practice this has meant that in traditional patriarchal liberal societies it has been women who, as supposedly 'innate carers', have been required to sacrifice their autonomy. By contrast, as he notes, the discussion of autonomy, pursuit of life-plans and so forth does not form a prominent element in the ethic of care.[32]

To this charge of neglect ecologists will wish to add another. In neglecting, or being simply ignorant of, the moral considerability of the non-human, traditional moral theorists have simply taken for granted that the needs of the non-human may simply be ignored in considering what is necessary to secure human autonomy. If recognition of human dependants makes the rhetoric of autonomy seem less relevant, so too does the recognition of the ways in which the pursuit of our own human projects, as individuals or groups, has to be circumscribed by the need to give proper recognition to the moral considerability of non-humans. Neither the ethic of justice nor the ethic of caring may be adequate to capture what is at issue in this dimension of moral thought.

Throughout the last four chapters we have focused primarily upon issues of morality as applied to politics. Occasionally we have touched upon the crucial issue of how the workings of the economy are affected by the implications of the basic value postulates of the modern political theories we have discussed, and of the distinctive basic postulate of ecologism. It is now time for us to consider such matters directly, which we will proceed to do in Part Four.

Notes

1. W. Kymlicka, *Contemporary Political Philosophy* (1990), p. 168.
2. D. Miller, *Market, State and Community* (1989), chapter 7 has a useful discussion of this.
3. Kymlicka 1990, p. 182.
4. See Karl Popper, *Objective Knowledge* (1972), p. 70.
5. Kymlicka 1990, p. 187.
6. Kymlicka 1990, p. 189.
7. Kymlicka 1990, p. 189.
8. For an example of such theorising see Robyn Eckersley, 'Green politics and the New Class: selfishness or virtue?' (1989).
9. Kymlicka 1990, pp. 188–91.
10. For specimens of communitarian writings see Alasdair MacIntyre, *After Virtue* (1981), Charles Taylor, *Hegel and Modern Society* (1979) and Michael Sandel, *Liberalism and the Limits of Justice* (1982).
11. Kymlicka 1990, pp. 202–3.
12. Kymlicka 1990, p. 206.
13. Kymlicka 1990, p. 210.
14. Kymlicka 1990, p. 212.
15. Cited by Kymlicka 1990 on p. 216.
16. Kymlicka 1990, p. 218.
17. Kymlicka 1990, p. 219.
18. Kymlicka 1990, pp. 223–4.
19. Kymlicka 1990, p. 227.
20. Kymlicka 1990, p. 229.
21. As cited by Kymlicka 1990 on p. 237, n13.
22. Kymlicka 1990, p. 246.
23. Such views are discussed by Plumwood 1993 and, under the heading 'cultural ecofeminism', by Dryzek 1997, pp. 158–9.
24. See Susan Moller Okin, *Women in Western Political Thought* (1992) for a thorough presentation of this argument.
25. Kymlicka 1990, p. 252.

26. Kymlicka 1990, p. 252.
27. Such as Carol Gilligan (1982), *In a Different Voice: Psychological Theory and Women's Development*, Cambridge: Harvard University Press, cited by Kymlicka 1990 on pp. 263–4.
28. Kymlicka 1990, pp. 264–5, 270.
29. Kymlicka 1990, p. 277.
30. Kymlicka 1990, p. 281.
31. Kymlicka 1990, p. 281.
32. Kymlicka 1990, p. 285.

Part Four

POLITICAL ECONOMY

10

Can Ecologism Transform Capitalism?: Sustainable Development, Ecological Modernisation and Economic Democracy

WHAT FORM AND LEVEL of economic activity is permissible if we wish to maintain biodiversity in the present world, thereby satisfying the requirements of ecological justice? This is a world in which there is a huge human population which is rapidly growing. It is one in which the most affluent societies, whose level of per capita wealth it appears to be the aspiration of the majority of the world's human inhabitants to attain, have only acquired that wealth by destroying habitats and reducing or exterminating many species. Does this not imply that, in the course of realising that aspiration, the teeming billions of humanity will necessarily destroy more habitats and exterminate many more species, at least up to the point at which the biosphere is so damaged that it is apparent to everyone that no further economic growth is possible without destroying the biological basis of human life itself?

The answer to this question which has been popular among many environmentally concerned people is that this is indeed what is implied if humanity follows its present course. Thus, if we are genuinely concerned to protect biodiversity, for all the reasons, prudential and moral, given earlier, then we have radically to alter our modes of economic activity as well as doing something rapid and effective to halt the escalation of the human population.

In this final part of the book we will be exploring some of the possibilities for the organisation of economic and political life which hold out some promise that human beings can acquire the material and other benefits necessary for their own flourishing in a way that respects the demands of ecological justice. However, this whole discussion is overshadowed by one inescapable fact, namely that the political economy of liberal capitalism, in spite of the vicissitudes with which it is endemically beset, is dominant, and is growing even more dominant, right across the globe.

There are good reasons, which we will shortly be going into, why political ecologists and other environmentally concerned people should regard this fact with serious concern. Their reactions to it divide roughly into two sorts. The pessimists see no real hope of altering capitalism's fundamental nature, and so see no real hope for an environmentally secure future. We are, so to speak, trapped aboard a runaway train which will stop only when it crashes into the environmental buffers.

The more optimistic adopt the approach of the exponent of Judo to the problem, seeking to use capitalism's own tendencies and momentum so to transform it that, in spite of its present tendencies, it will at least begin to conform to the requirements of ecological health and justice. Capitalism may never satisfy these requirements to any great degree, but this approach hopes at least to buy time for an alternative to capitalism to be found.

Political Economy: Critique and Counter-critique

Let us first survey the general debate between environmentally concerned people and their critics concerning matters of political economy in order to set the stage for the rest of the discussion. There are two distinct issues involved in this area. The first is the question of the material viability of our current economic practices, which in turn encompasses two general problems. These are:

(1) Resource problems – will we have the natural materials to sustain the economic activities of a rapidly growing population? 'Natural materials' encompasses land, water, fuels, ores and so on.
(2) Sink problems – will we be able to dispose of the waste products of our economic production in a way which does not poison or destroy our sources of clean water, land and air?

The second issue concerns the human significance of our economic activity. For we do not simply produce in order to meet physical needs. In the course of material production, or praxis, we transform our lives and ourselves in significant ways. Human beings exercise their imaginations, their emotions, their moral sense and their religious beliefs in the course of material production. In identifying and overcoming the problems of material production by the invention of new technologies and ways of working, human beings have come to make over the world in a way which renders it more humanly intelligible. On some views, such as those of Hegel, they have, on balance, come to be more 'at home' in the world as a result. If this is a correct analysis, then the question raised by proposals to alter the basis of human economic activity is whether they amount to a viable praxis.

To this list of issues the ecologist adds one more. This is the issue of whether, if we can find a humanly satisfying form of economic activity which avoids or minimises resource and sink problems and deals with the problem of human poverty, it will also satisfy the requirements of ecological justice.

Problems (1) and (2) were among the first to be recognised and highlighted by environmentalists. An alarming picture emerged from their analyses of human beings as about to run out of material resources and to pollute the planet to a terminal degree. This picture underpinned appeals addressed directly to human self-interest and led to the idea that the system of economic production to which we are apparently committed, far from being conducive to human well-being, would lead in a short time to human misery of untold proportions.

From this diagnosis have emerged well-known environmental themes:

(1) The critique of our present conception of economic growth, based on Gross National Product (GNP) and Gross Domestic Product (GDP), which measures as a positive contribution to human well-being all the measures we have to take to remedy self-inflicted harms, arising from the profligate and inequitable forms of economic activity in which we engage.[1]
(2) The injunction to cut down on waste, to recycle as much as we can of the materials used in manufacture, and the emphasis on the need to turn to renewable sources of energy and to reduce our consumption of non-renewable fossil fuels and ecologically suspect nuclear power.
(3) The attack on the increase in polluting and greenhouse gases, such as carbon dioxide, released by the use of fossil fuels in the course of our modern manufacturing industries and modes of

transportation, coupled with the highlighting of global warming and ozone depletion as examples of harmful large-scale human effects on the biosphere.

(4) The attack on the polluting, ecologically harmful and inhumane methods of modern mass-production agriculture, and the urging of farmers to turn to organic, humane and ecologically beneficial methods for growing our food.

(5) The claim that human numbers and economic activity adversely affect habitats and biodiversity. This adverse impact is said to be substantial purely in terms of what it means for human economic resources, ecological services and the conditions needed for human culture to flourish, even without going into the claims political ecologists also wish to make about the moral considerability of the non-human.

(6) The claim that the enormous increase in human numbers which threatens to enhance the environmentally harmful impact of the contemporary economic system to a disastrous extent is at least partly caused by poverty, which has to be dealt with by a more equal distribution of resources rather than by extension of the pernicious forms of economic activity among the poverty populations of the world.

These claims have all been subject to vigorous counter-critiques by defenders of the existing economic system and economic growth as understood within orthodox economics. Before we look at these criticisms we need first to present, from the environmentalist perspective, the possible causes of the above phenomena.

Controversy has arisen among environmentalists on this matter. Some lay the emphasis on the rise of modern industrialism as the prime culprit in the creation of an ecologically harmful economic system. This ushered in the era of mass-production with its concomitant demand for the extraction and use of enormous amounts of raw materials and fossil fuels. On this view, industrialism, wherever it occurs – under a capitalist, socialist or any other system – is what has to be radically restructured in accordance with ecological imperatives. Fitting in with this view there is the call for the adoption of simpler, less materialistic lifestyles and the turning to more spiritually rewarding activities which involve harmony with, rather than destruction of, the environment.[2]

Others, such as David Goldblatt, regard this 'industrialism-plus-materialistic-greed' analysis as being too superficial, and point to the inner dynamic of capitalism, with its ceaseless quest for profit and concomitant urge to create ever new markets and endlessly to stimulate human demands for material goods and services, as the main culprit.[3] To the obvious criticism of this view that the worst environmental records this century are to be found within those societies which abolished capitalism and sought to employ a system of rational planning of economic activity to meet human need directly, Goldblatt replies that such societies were seeking to emulate the practice of capitalism without the democratic safeguards present in capitalist societies. The latter, he argues, have mitigated some of the worst effects of capitalism precisely by permitting the emergence of an environmentally conscious social movement.[4]

To this is added the claim that industrialism on its own would not have had the environmentally harmful effects with which we have become familiar in the late twentieth century. As Goldblatt argues, only when industrialism was organised within the capitalist system which it had itself unleashed did the potential created by the discovery of coal-based forms of energy-production to release economic activity from previous ecological limits become a reality.[5]

From this perspective the aim of environmentally concerned people should be to control rigorously the workings of capitalism in the interests of the preservation of the biosphere. This will involve a mixture of state and international regulation, taxation policies, licensing regimes and the increased democratic accountability of capitalist corporations. It may also involve encouraging capitalist firms to engage in the activities which go under the heading of 'ecological modernisation'. This involves, among other things, the manufacture of environmentally beneficial technologies – for example, catalytic converters and other devices to reduce the environmentally harmful effects of economic production and transportation – by processes which are themselves made minimally damaging to the environment.

Having outlined the environmentalist perspectives on matters of political economy, let us now rehearse the counter-criticisms of all that has been said so far offered by those who are sometimes labelled 'contrarians'.[6]

The basic strategy of critics of the environmental views listed above is to claim that where there are genuine problems they can be, and are being, satisfactorily tackled within the economic dispensation as it exists in the late twentieth century, but that most of the 'problems' are in reality non-existent (or, more weakly, that the case that there is a genuine problem has not yet been established). Thus, the resource problems are being dealt with by the usual means of finding alternative raw materials and sources of energy. Human ingenuity is the key to their solution. Solar and other forms of renewable energy, new forms of nuclear energy without the creation of nuclear waste, even the long-sought fusion power, when coupled with already demonstrated ability to use energy more efficiently, will cope with any energy needs we are likely to have, even with a greatly enlarged human population.

New materials and production techniques will cope with any potential resource shortages. Human ingenuity is the crucial resource here once again. Thus, the possibility of nanotechnology, which involves the creation of any material we desire from atoms, promises to cure whatever raw material shortages may threaten, and key steps have already been taken towards it. In any case, much modern technology is based on the computer, the production of which involves a reduction in the amount of material used concomitantly with massive increases in computing power achieved. Further, much economic growth in the future will come from forms of activity which are much less material- and energy-intensive than the smokestack industries of the early phases of the Industrial Revolution. Information and education are the key to future economic growth, and these use far fewer material resources than their predecessors.

In the area of pollution and waste, critics of environmentalism point to the way in which the pollution-creating activities of units such as factories, power-stations and automobiles have all been drastically reduced in recent years, partly as a result of the recognition that polluting forms of production are usually inefficient forms of production. They emphasise the robustness of the biosphere and its ability to recover very quickly from mishaps (miscalled 'disasters' by environmentalists) such as large oil spills at sea. They point to the way in which alternatives were found to the ozone-destroying CFCs and confidently expect that the ozone layer will rapidly repair itself, or at least not deteriorate further – although this is a problem which they claim was overblown in the first place. They strongly challenge the claim that global warming is being produced as the result of releases of greenhouse gases into the atmosphere in the course of human activities, but argue that, even if it is occurring, there is no reason to suppose that its effects will be sufficiently harmful to warrant strenuous steps to prevent it. In their view we are ignorant of what its effects are likely to be, and for all we do know it is as likely to have beneficial effects overall as it is to be harmful.

In the sphere of agriculture they argue that the problems of pesticide poisoning of the environment have largely been overcome by the invention of transient and more finely tuned pesticides. Genetically engineered plant strains also promise to make the application of pesticides less extensive. They simply reject organic agriculture as a romantic fad, and point to the enormous increases in agricultural productivity which science and fossil-fuel-based agriculture have achieved and which further developments in biotechnology promise to increase in the future. Thus, they argue, there is no reason why the whole of the expected increase in humanity should not be adequately, or even well, fed in the future, in the manner to which the fortunate citizens of the West have become accustomed since the early years of the twentieth century, provided that the advances possible under scientific agriculture married to the market economy are not stymied in the name of a misconceived concern to protect the environment.

With respect to the decline in habitats and the loss of biodiversity, critics are again inclined strongly to be sceptical of the claim that serious losses of species are occurring, or are likely to occur.[7] Even if the claims of environmentalists are correct in this regard, critics of environmental arguments challenge the view that any serious harm to the biosphere will ensue, or that human beings will forego large economic, scientific and medical benefits potentially present in the lost species. They further claim that it is not proven that reduction or fragmentation of habitats result in extinctions – or in large-scale extinctions, at any rate. In any case, humans come first, they argue, and if the cost of creating a decent life for human beings is the loss of some species and habitats, then this is a price well worth paying.

The moral arguments of ecologists are, of course, rejected. Non-humans are held to have only instrumental value, in so far as they are necessary for humanity's economic, aesthetic or, at a pinch, spiritual benefit. The view that non-human creatures are morally considerable and intrinsically valuable is regarded as an affectation of the comfortable Western middle class, reacting to the disenchantment of the world resulting from the success of modern science and their loss of religious faith. It is not to be taken seriously.

Live for a while as a Third World peasant, critics say, facing the perpetual threat of famine, drought, earthquakes, floods, disease and the depredations of wild animals, from bugs to elephants, and the truth about the non-human natural world will quickly become apparent, namely that it is a constant threat to human happiness and well-being. Those human beings may count themselves fortunate whose forebears, by their enormous economic and technological efforts, released their descendants from thraldom to the natural world, allowing them to dwell on its more aesthetically pleasing aspects from the comfort of their totally artifical urban or rural environments based on the subjugation of the natural world to human will. This is the condition to which all human beings, rightly, aspire.

These ripostes to the environmentalists' claims are underpinned by the critics' celebration of economic growth as hitherto understood, and by their emphasis on the vital instrumental importance of capitalism, free trade and the market economy in generating and sustaining that growth. At which point we move on to the more familiar and much-debated issues of the merits of capitalist markets and their relationship to issues of justice, equality and liberty among human beings, an issue which, as we have seen, is of direct interest to environmentalists, but not only to them.

It is clear that this series of disputes between environmentalists and their critics covers issues which are partly empirical and partly concern rival value-systems. On the empirical side, the critics are on their strongest ground in their claim that resource problems have been misconceived or exaggerated by environmentalists. Here, human ingenuity is indeed still promising as the key to finding the energy sources and raw materials needed to sustain our material culture.

However, they become less and less convincing as they venture further into fields in which the health of the biosphere is in question. They appear overly sanguine about the possibilities of mass, fossil-fuel-based agriculture, whose monocultural basis remains a hostage to fortune. Their confidence that the problems associated with pesticides have now been overcome are premature, to judge by a 1997 UK report on the harmful indirect effect of temporary, targeted pesticides on British birdlife.[8]

They are correct to note that there have been important steps forward in making units of manufacture and transport less polluting, at least in the West, but what they have not shown is that the benefits of reducing unit pollution will not be swamped by the enormous increase in the sheer number of production and transport units, as the Third World catches up with Western patterns of industrialisation.

They seem to be correct in their view that the depletion of the ozone layer, if it is indeed due to the emission of CFCs, is a problem which is easily curable, at least in principle. But they are beginning to look somewhat cavalier in their nonchalance with respect to the isue of global warming. It is clearly correct that we do not know what its effects, if any, will be, but it is not unreasonable to suppose that it is occurring, that human activity is involved in its production, that its effects will be very marked, and that it is at least possible that those effects will be far from beneficial. At the very least this looks to be a game of Russian roulette on a massive scale.

They appear to be even more cavalier in their attitude towards biodiversity and habitat destruction. The loss of species referred to by ecologists is not going to result in neat piles of visible corpses for human beings to count easily. The species, especially those located in the all-important rainforest habits, are usually going to be of creatures which are small in size and highly localised. Extinction, even when rapid, need not produce any dramatic visible effects, even for experienced and skilled observers. As ecologists dramatically but correctly point out, the species thereby lost are lost forever. If they disappear before we even find out that they exist then we have no way of even contemplating resurrecting them by the wizardry of biotechnology.

It is correct to point out, as the critics of environmentalism often do, that the evidence shows that, even with dramatic pollution incidents such as massive oil spills, natural processes, when allowed to operate, can bring the affected environment back to something like its previous condition quite quickly. Nature is indeed very robust, as the history of extinction episodes during the last 600 million years has revealed. But, although it is difficult to damage the biosphere beyond repair this does not imply that it is impossible to do so. Although it would be hard for human beings to create as much havoc on the biosphere as was produced by the asteroid whose collision with earth exterminated the dinosaurs and enormous numbers of other life-forms, an all-out nuclear war among human beings would not be a bad approximation to it.

What this suggests is that since the 1940s human beings have acquired the possibly unique power to inflict upon the biosphere serious, perhaps terminal, damage. Given our enormous appetite for materials, the huge and rapidly increasing size of our population, unprecedented for a mammal of our dimensions, and our ignorance concerning the precise effects of our activities upon the biosphere, it ill becomes any of us to be other than extremely cautious in our estimate of whether or not the biosphere can take all we can throw at it.

However, the critics clearly have a good point in their emphasis on the fact that for most human beings, for most of their history, the biosphere, besides providing the conditions to sustain humanity in existence, has also constantly threatened it with a whole host of unpleasant phenomena. The protection of ourselves from these threats correctly remains a main priority for human activity. But this has to be balanced against the fact that even Third World peasants show a love of their natural environment, for all its actual and potential harshness, and that in the practical business of conservation there are many people in the Third World who regard it as important, not just for economic reasons. Some such people devote their lives, often at risk to themselves, to its protection.

There exist many Third World environmental protection organisations, often with members drawn from a peasantry which sees its immediate natural environment as under threat from the rapacity and/or thoughtlessness of governments, corporations and powerful individuals. Thus, it is just not true that concern for the welfare of non-humans is merely a Western middle-class fad. To see this one has only to recollect that many religious and moral viewpoints pervasive in the Third World and among aboriginal people in general testify to the importance they attach to a proper respect for our fellow-creatures.

One might also add that the threats posed by the natural world are at least matched by the threats which human beings pose to each other in the form of war and genocide, and that often the ill-effects of the non-human world upon human beings is mediated by human malice and stupidity – consider how often famines have occurred in situations where food was in plentiful supply but witheld from those facing starvation.

As far as the attack on the moral views of ecologists is concerned, all that can be done is to pursue the arguments, such as those presented in Parts Two and Three of this book. These may be faulty, but, if they are, that has to be shown to be so by the normal procedures of rational debate, not by the *ad hominem* condemnation of them as the sentimental conceits of middle-class Westerners with nothing better to think about.

Versions of Sustainable Development

Having argued that the environmentalists are correct to be concerned about the workings of untrammelled capitalism, we need to contemplate some of the ways in which capitalism may be redeemed, environmentally speaking. Three pertinent ideas require discussion here, namely sustainable development, ecological modernisation and economic democracy. In the next chapter we will consider two alternatives to capitalism in any of its possible forms, namely market socialism and global ecology.

It is by now standard to note that there has been developing since the publication of the Brundtland report in the late 1980s the idea that it may be feasible to attain the economic growth necessary to deal with widespread human poverty in a way which does not involve sawing off the biospherical branch upon which we are sitting.[9] Humanity, it is claimed, has the possibility of instituting a process of 'sustainable development', which involves so acting that we bequeath to our descendants a world which is as good as the world which we have inherited from our forebears.

Whether or not the concept of sustainable development does make sense, it is a tribute to the influence of environmental thinkers that it has been thought necessary in a major United Nations report to try to rethink the basis upon which economic activity is carried out. However, we need to consider whether the idea of sustainable development in some form can commend itself to ecologism, or whether instead political ecologists have to propose a radical abandonment of traditional economic goals. This could encompass the institution of a much lower level of economic activity together with a much simpler and less affluent lifestyle, at least in material terms, and a greater willingness to share what we do possess in the way of material resources.

As Munda has pointed out, we are faced with at least two conceptions of what sustainable development actually is.[10] The view prevalent in the sub-discipline of neo-classical economics called 'environmental economics' is that we should understand sustainable development in accordance with the criterion of 'weak sustainability'. To use Munda's formulation, an economy can be considered weakly sustainable if it saves more than the combined depreciation of natural and man-made capital. As Turner et al. put it:

> We can pass on less environment so long as we offset this loss by

increasing the stock of roads and machinery, or other man-made (physical) capital. Alternatively, we can have fewer roads and factories as long as we compensate by having more wetlands or mixed woodlands, or more education.[11]

The main objection given by Munda to this conception is that it assumes the perfect substitutability of different forms of capital, natural and man-made. This concept has to be rejected because, as he explains:

(1) Man-made capital is not independent of natural capital: in creating capital, natural resources must be used, so that the attempt to substitute such artificial capital for natural capital will be limited by the extent to which the increase in artificial capital requires an input of natural resources. The only way to avoid this being a problem is to create capital whose productivity outweighs the natural resources used up in its creation, and it cannot just be assumed that this is possible.
(2) Artificial capital is monofunctional (a road exists solely to carry traffic) whereas natural capital is multifunctional (woodland provides food and shelter, acts as a sink for carbon dioxide, is a source of wood).[12]

To this has to be added the important consideration adduced by two economists working within environmental economics, David Pearce and Kerry Turner, who argue that the maintenance of the natural capital stock is a condition for sustainable development. This is because we do not possess full scientific knowledge of how the biosphere works nor of its precise role in supporting human economic systems, thus we must retain it as far as possible, on pain of destroying irreversibly something which may turn out to be essential to our economic and other well-being.[13]

An alternative version of sustainable development is that put forward by the newly developing discipline of ecological economics. This rejects a whole host of the assumptions present in neo-classical environmental economics, such as the possibility of calculating costs and benefits of different economic activities by reducing all factors to a single criterion of comparability, namely the monetary one. Ecological economics argues that sustainable development must employ the criterion of strong sustainability. As Munda characterises it, this assumes that certain kinds of natural capital are critical to the health of the biosphere and that such capital is not substitutable by means of man-made capital. Instead, we need to develop nonmonetary indicators of ecological sustainability, based on the direct physical measurement of important stocks and flows of natural capital.

This will be inherently problematic for many reasons, among which are the fact that such indicators will be subject to scientific uncertainty and to political negotiation, as different actors attempt to persuade decision-makers to adopt their preferred interpretations. Thus, the working out of what is to be done to maintain economic activity across the globe in a strongly sustainable mode requires the development of democratic processes for the articulation and negotiation of incommensurable values and considerations.

If we ask the question which version of sustainable development ecologism should support, it is clear that it has to be that espoused by ecological economics. Defenders of biodiversity on the ground of the intrinsic value of species are *ipso facto* committed to the rejection of the valuing of species in purely monetary terms, that is, they are committed to rejecting the view of species as purely 'natural capital', even though they accept that each species is also potentially a resource for human economic purposes.

Hence, they must be attracted to a view of economic decision-making which allows for the fact that different, mutually incompatible, and irreducible values are legitimately in play. They will also be happiest with a view of sustainable development which emphasises the need to maintain natural capital, for that promises that the case will be more readily acceptable for the preservation of natural habitats within which biodiversity may be preserved.

Finally, the emphasis which ecological economics places on the centrality of democratic processes of decision-making is one which fits into ecologism's case for democratic decision-making and the guardianship institution outlined in the previous part of this book. The latter will be a key player in the democratic debates which are envisaged as being central to the decision-making process, even if, as we saw in Chapter 8, there are good reasons for insulating the members of such an institution from some of the direct effects of democratic politics.

However, clearly there are great difficulties involved in working out methods for finding indicators of key elements in the natural environment in order to conduct the debates. It is also clear that ecological economics does not seem to have committed itself to the idea that any group participating within the debate will be accorded a veto power, whereas that is something envisaged by the creation of the guardianship institution. However, it is part of the logic of the non-substitutability of different forms of capital that someone should be able to block proposals to destroy a critical part of the natural capital, especially as such decisions are nearly always non-reversible. Hence, provided that there is scientific knowledge to sustain it, a veto over the destruction of key elements of natural capital is necessary.

A crucial element in the viability of democratic decision-making which aims at sustainability is, as has forcefully been argued by Gray, that the accounting procedures of capitalist corporations must be radically altered to make transparent to democratic decision-makers what precise impact the economic activities of such corporations are having upon the sustainability of the human economy.[14] He argues, with great plausibility, that neither voluntary actions by corporations nor the workings of markets will move far enough and quickly enough in the direction of what he calls 'reporting for sustainability'. At the end of the twentieth century, both markets and corporations respond solely to financial indicators. The fact that their impacts on environmental processes often resist expression in financial terms requires that new forms of reporting of their activities be devised and enforced.[15] Even when these are in place, however, capitalist firms will not be in a position themselves to make informed decisions about the sustainability of their activities, nor is it reasonable to require them to take the sole responsibility for acting sustainably. This is a matter for democratic decision-making by the wider society, involving, as it necessarily does, judgements about trade-offs involving a whole variety of factors.

Thus far, the concept of sustainable development, although of considerable interest to ecologism, remains at a rather high level of abstraction. A more specific strategy for 'greening' political economy, specifically in its capitalist form, is that known as ecological modernisation.

Ecological Modernisation

The concept of ecological modernisation is simple enough to explain. In Dryzek's formulation it involves the restructuring of the capitalist political economy along more environmentally sound lines.[16] Adverse environmental effects are held to stem from the way that productive and distributional processes are organised within late twentieth-century capitalist economies. No inherent difficulties are believed to stem from the capitalist form itself. The trick is, therefore, to induce the appropriate forms of economic reorganisation within capitalist enterprises in a way which protects profitability while reducing or minimising adverse environmental effects, such as pollution.

Dryzek argues that the capitalist economies which have been most successful so far in this endeavour have been those of Germany, Holland, Japan, Sweden and Norway. They have been conspicuously successful in increasing the energy efficiency of national income (by, for example, changing to more energy-efficient processes), reducing the amount of harmful emissions caused during economic activity and reducing the amount of waste produced.[17] These countries, partly as a result of their corporatist political culture, have achieved consensus and co-operation between government and capitalist firms on the appropriate measures. Most obvious in this regard has been the creation of the Dutch National Environment Policy Plan, adopted in 1989.

In part this has also been the result of the relative willingness of capitalist enterprises in these countries to take a longer-term view of what will lead to economic viability than the annual statement of profit provides. Hence, in Germany the 'precautionary principle' has now gained widespread acceptance, namely that the absence of scientific certainty as to the nature and degree of an environmental threat does not justify doing nothing to tackle it. This is an exercise in foresighted prudence of a kind often held to be incompatible with the short-termism of capitalist practice.

However, as Dryzek also points out, the securing of a profit has not disappeared as the main capitalist aim. Rather, ecological modernisation is held to be good for business. He cites five main advantages for capitalist firms in climbing aboard this bandwagon. Less environmentally harmful forms of production are less inefficient; tackling problems at an early stage obviates worse costs later; employees are happier and work better in good environments; there is money to be made in selling goods and services with 'Green' credentials to an increasingly environment-conscious market; profits can be made from producing and selling products designed to reduce or prevent pollution.[18] One might summarise this by saying that in these areas the conventional methods of accounting, attending solely to factors which can be given the usual financial expression, point capitalist corporations in a more environmentally beneficent direction.

Several important points for ecologism emerge from this phenomenon. First, it provides an example of capitalist enterprises apparently being willing to accept that governments have a legitimate role in trying to alter the organisation of capitalist firms in the name of the wider public good. Secondly, it shows that the long-term view is not hopelessly out of phase with capitalist concerns. Thirdly, it shows that the holistic, systems-oriented approach of ecology, and ecologism, concerning interactions between humanity and the rest of nature can be accepted by both capitalist enterprises and governments. Finally, it shows that even if the replacement of the market system by a fully planned economy is no longer a viable alternative, there is scope for planning in the sphere of economy–environment relations. The Dutch National Environment Policy Plan set environmental quality targets and a timescale within which they were to be achieved, based on a theoretical understanding of the processes involved.[19]

All of these elements suggest that ecological modernisation is something which at least reveals the potential for a form of political economy within which ecological justice can be found a place – although, of course, this is still a long way off. The limitation of the scope of democracy represented by the guardianship institutions also looks less at variance with the capitalist system if the latter is itself willing to accept some legitimate government intervention with respect to its often trumpeted 'freedoms'.

However, ecological modernisation looks at the moment like a game for rich, successful capitalist countries who are willing to countenance some technological tinkering with their economic processes to achieve some clear-cut, if limited, environmental improvements within their borders, provided that profits and competitiveness are not compromised. But none of this may be affordable elsewhere, and the environmental injustices involved in the export of toxic waste, the expropriation of raw materials at low rates of remuneration from the Third World and so forth, may mean that this phenomenon is largely window-dressing.

Nevertheless, as Dryzek notes, these developments may open the door to a much more radical ecological restructuring of capitalism.[20] Following Christoff and Hajer, Dryzek moots a development of ecological modernisation along the lines of the institutionalisation of democratic participation in the environmental planning of whole societies in a way which also encompasses the international dimension.[21] At this point, then, in its turning to the need for wider democratic decision-making, the discussion of ecological modernisation joins up with the discussion of sustainable development in the last section. In the longer term, Dryzek suggests, this process might even lead to the 'exit from industrial society'.[22] This might emerge as a result of the effect of what Beck has dubbed 'the risk society', which is one which generates severe environmental risks with which it has no effective way to deal.[23]

However, we are still clearly embedded in an industrial society which, under the impulse of capitalism, is spreading to newer areas of the planet with, as yet, no serious impediment. It is worth noting at this point the features of capitalism which, in spite of the promise of ecological modernisation, make it problematical from the point of view of ecologism.[24]

Capitalism remains resolutely human-chauvinist in its orientation. Human welfare and/or preference-satisfaction is the sole ethically important demand which it can admit. Thus, environmental issues are couched either in terms of prudence (what is instrumental for human welfare) or in human 'quality of life' terms. The idea that the non-human may possess moral considerability which circumscribes human economic activity is one which capitalism as such cannot encompass. This is because capitalism is designed only to respond to human preferences expressed in the marketplace. Although individuals who are capitalists are capable of wider moral concerns than this, *qua* capitalists they can only be concerned about market-expressed preferences.

However, capitalism's responsiveness even to human preferences is contingent on their being expressed in the market-place. It cannot respond directly to human preferences not so expressed. This in turn means that, for capitalism to respond to non-human preferences and needs, the latter must first be reformulated as human preferences (for example, that certain species should continue to exist) and these must then be made manifest to capitalist enterprises by either or both of consumer preferences (for green products/services) and politically created regulations and 'market' mechanisms (taxes and subsidies). Hence, capitalism *qua* capitalism has no intrinsic tendency to show any responsiveness to non-human interests (or to many human interests either).

By contrast, the profit motive is central to capitalism. This fact alone makes the ideas of 'stakeholding' and 'ethical investment' hard to embed within capitalism, for they have the clear tendency to compete with that motive, and it is a motive which has to be given total allegience by capitalists, on pain of failing in their capitalist enterprise. As we suggested above, proposals put to capitalists for greater environmental reponsibility have to be shown to be commercially viable if they are to be given a hearing. Further, if consumer preferences and government regulatory and 'market' activity threaten capitalist firms' profitability the latter will have a very strong tendency to try to mould or manipulate the consumer preferences and to influence politically the governmental actions so that opportunities for profit are not diminished.

The profit motive is not eliminable from capitalism, nor is the commitment to economic growth. Jackson has given a clear explanation of how these two phenomena are connected via the key ingredient of capitalist financial markets, namely, the charging of interest on the loans necessary for capital investment. Roughly, there can be no investment without loans, no loans without interest being charged, no payment of interest without the making of a profit and no maintenance of production (and thus securing of a livelihood) without further investment to be paid for by further loans, more interest and increased output and/or lower costs to repay them, all of which adds up to economic growth.[25] The importance of this argument is that it points to forces within the structure of capitalism to explain why growth is an ineradicable feature of it. It does not rest on an analysis of capitalists as greedy or avaricious, even if some are and are attracted to a capitalist life because they are. Thus the cure for growth fixation will not come about as the result of making sure that capitalists are not subject to such vices.

An important threat to capitalist profitability comes, of course, from competitors in the market-place. The competitiveness is crucial in stimulating efficiency, technological innovation and creating consumer power, which are held to be key benefits of the market system, heightened in its capitalist form by the private ownership of the means of production. But this need to compete is what provides a standing temptation to capitalism to cut corners, environmental or otherwise, to maintain or gain a competitive advantage.

If capitalism is necessarily committed to profitability, and if profitability requires growth in output, then capitalism is necessarily committed to growth. There are, as we have seen, ways of making economic growth less materially intensive, polluting and wasteful, but the impact on the environment in terms of use of resources and pollution cannot be reduced to zero. With constant or increasing rates of growth at some point the growth curve starts to rise steeply (the well-known phenomenon of exponential growth). We can defer this by reducing the rate of growth, but unless we can keep reducing the latter without limit, at some point the steep rise must kick in. Hence, the best we can do with capitalism is to reduce its environmental impact for a time until, as Jackson suggests, we find a different political economy to practise.[26]

During this period we may hope that the one environmentally beneficial development internal to capitalism will continue to pay dividends, namely the discovery that making production less polluting and more parsimonious in use of material and energy cuts costs and aids profitability. But the points noted above entail that the main forces keeping capitalism pointing in a greenish direction must come from outside it – consumer pressure and governmental regulation and market activity.

Ecologism should conclude, therefore, that while much can be done immediately to make capitalism more environmentally benign, and economic growth, capitalist or otherwise, will be needed for some time to produce improvements in human well-being in the poorest countries, we do need to look beyond growth-oriented economic systems in general, and capitalist ones in particular, if we are to have a political economy which secures both human well-being and allows for the moral considerability of the non-human.

What would such a system be like? We can say at this stage only that it would be one in which the material needs of a stable human population would be met in a manner which minimised material throughput and human beings' primary concern would switch from the accumulation of material wealth to non-material needs, such as affection, creativity, the contemplation of beauty and the pleasures of human intercourse (presumably themselves satisfied on a largely non-material basis).[27]

Economic Democracy

Let us finally survey the possibilities and problems, from the point of view of ecologism, of democratising the internal structure of capitalist firms. This possibility, called 'economic democracy', has recently been discussed by Archer.[28] The case for such democracy is derived by Archer from the basic value to be attributed to autonomy for individual human beings, understood as making one's own choices about how to act. The case for autonomy in joint decision-making leads to the case for democracy in general and to two specific principles:

(1) The 'all affected principle' – all individuals whose ability to make choices and act upon them is affected by the decisions of an association (such as a capitalist firm) should share control over the process by which those decisions are made.[29]
(2) The 'all subjected principle' – all individuals who are subjected to the authority of an association should have direct control over the decisions of that association.[30]

Principle (1) applies to the 'stakeholders' in a capitalist enterprise, comprising employees, consumers, shareholders, suppliers of raw materials and producer goods, financial institutions and local/non-local residents whose lives are affected by the firm's externalities. Such stakeholders' control has to be indirect, that is does not involve participating in the internal decisions of the firm. The two indirect methods of control available to stakeholders are 'exit', or removing oneself from relations with the firm as worker, consumer, shareholder and so on, and government regulation, especially where no exit is available as a means of exercising control.

Principle (2) applies to the employees who are, under capitalism, subject to the authority of the firm which employs them. That is, *qua* employees their actions are governed by the choices of the firm's managers, not by their own choices. It is with respect to them that principle (2) enjoins the creation of direct decision-making within the enterprise. To shareholders would remain the exit strategy of control which is, Archer claims, what in most cases is the main instrument of shareholder control even under present circumstances.

The question of whether such decision-making would impair the economic efficiency of capitalist firms is, Archer argues, an open question. He notes, however, that if it could be conclusively demonstrated that economic democracy was bound to be much less efficient than present forms of capitalism then its acceptability would have to be seriously questioned.[31] This is intelligible on the premise of autonomy, for an economy which was operating inefficiently would itself be detrimental to the exercise of human autonomy by depriving people of some possibilities of choice. However, he notes that the question of the viability of a system is value-laden and depends upon what we demand of it. If economic efficiency is not our sole goal, then we may reasonably accept a less efficient economic system in order to achieve some other important goal.[32]

From the point of view of ecologism, the autonomy-based case for economic democracy is an attractive one, as already noted. However, one element in Archer's analysis is obviously open to criticism from this point of view, for it can be said that missing from the list of stakeholders, or those individuals affected by the decisions of associations, are non-human living beings whose well-being may well be adversely affected by associations' activities. For these stakeholders, exit is not possible as a means of control, because they are not in conscious relationships with human associations, and so cannot employ exit to assert such control. Some animals may be able to flee adverse effects, but, even if this, for some reason, inflicts a penalty on the firm, it still does not amount to control. Nor, as non-participants in politics, can they directly obtain government regulation to protect their interests.

Of course, the benefits of both exit and regulation are available to them indirectly, provided that human consumers and shareholders withdraw from relations with firms which ignore requirements of ecological justice, and elected politicians enact regulations with the support of concerned citizens to enforce the requirements of such justice.

What this highlights is that, from the point of view of ecologism, economic democracy's prime justification will be on the basis of human autonomy, but that it will require the practices of economic democracy to be subject to external forms of control, especially from the guardianship institutions at all levels, aiming to ensure that ecological justice requirements are met. There is nothing inherent in the idea of any kind of democracy which guarantees that ecological (or environmental) justice is respected. This applies also to economic democracy.

Having surveyed the reasons why ecologism should be prepared to offer an extensive critique of the currently dominant form of political economy, and having surveyed the reasons for both optimism and caution concerning the possibility of transforming it in ecologically benign directions, we need finally to address arguments coming from the other end of the current debate. One recent Green discussion of this kind involves a thorough-going scepticism directed towards the rhetoric of sustainable development and attempts to produce a version of political economy which is very unlike any form of contemporary capitalism. We should first, however, consider an ingenious alternative to capitalism emanating from the socialist, rather than the Green, camp and assess it from the point of view of ecologism.

Notes

1. The critique of the neo-classical tradition in economics and its offshoot, environmental economics, focusing on GNP/GDP measures of growth, cost-benefit analysis, shadow pricing and so forth may be perused in such works as Herman Daly and John Cobb, *For the Common Good* (1990); Michael Jacobs, *The Green Economy* (1991); Frank Dietz and Jan van der Straaten, 'Economic theories and the necessary integration of ecological insights' (1993).
2. In this vein, see Keekok Lee, 'To de-industrialise – is it so irrational?' (1993).
3. See David Goldblatt, *Social Theory and the Environment* (1996), pp. 34–8.
4. See Goldblatt 1996, p. 49.
5. Goldblatt 1996, p. 35.
6. In what follows I am lumping together some theorists and theories which have been illuminatingly teased apart by John Dryzek in *The Politics of the Earth* (1997). He distinguishes between Prometheanism and Economic Rationalism, both hostile to any radical environmental discourse and both pinning their hopes for the continued material progress of humanity on a combination of markets, private property, self-interest and human ingenuity. Their chief difference concerns the rejection of any limits to growth by the Prometheans (who are thus also Cornucopians) as compared with the willingness of Economic Rationalists at least to contemplate the possibility of such limits. For my purposes a more broad-brush approach to the depiction of the critics of environmentalism is all that is needed. Works such as Wilfred Beckerman's *In Defence of Economic Growth* (1974) and Herman Kahn's and Julian Simon's *The Resourceful Earth* (1984) are classic specimens of the genre.
7. See Greg Easterbrook, *A Moment on Earth* (1996) for a recent example of this scepticism.
8. Department of the Environment, *The Indirect Effects of Pesticides on Birds*, London: HMSO, 1997.
9. See World Commission on Environment and Development, *Our Common Future*, 1987.
10. Giuseppe Munda, 'Environmental economics, ecological economics and the concept of sustainable development' (1997), p. 215. In this section I am relying heavily on this very instructive article.

11. Turner et al., *Environmental Economics: An Elementary Introduction* (1994), p. 56, quoted by Munda 1997, on pp. 217–18.
12. Munda 1997, p. 217.
13. This argument is presented by Munda 1997 on p. 218, citing D. W. Pearce and K. R. Turner, *Economics of Natural Resources and the Environment* (1990).
14. R. H. Gray, 'Corporate reporting for sustainable development', 1994, pp. 17–45.
15. He explores three possible ways of doing this – the Inventory Approach; the Sustainable Cost Approach and the Resource Flow/Input–Output Approach (pp. 32–7).
16. Dryzek 1997, p. 141. Albert Weale, *The New Politics of Pollution* (1992) is a valuable study of ecological modernisation.
17. Dryzek 1997, p. 137.
18. Dryzek 1997, p. 142.
19. Dryzek 1997, p. 138.
20. Dryzek 1997, p. 145.
21. Dryzek 1997, p. 148.
22. Dryzek 1997, p. 149.
23. Ulrich Beck, *The Risk Society* (1992).
24. A positive assessment of capitalism's environmentally sound prospects may be found in John Elkington and Tom Bourke, *The Green Capitalists* (1989).
25. Tim Jackson, *Material Concerns* (1996), pp. 167–8.
26. See Jackson 1996, p. 176.
27. See Jackson 1996, chapter 7, for interesting suggestions concerning how one might replace material goods with the services those goods are designed to provide.
28. Robin Archer, *Economic Democracy* (1995).
29. Archer, 1995, p. 27.
30. Archer, 1995, p. 32.
31. Archer, 1995, p. 63.
32. Archer, 1995, p. 63.

11

Alternatives to 'Greened' Capitalism: Market Socialism and Global Ecology

Market Socialism

THE MOST THOROUGH RECENT case for market socialism is that put forward by David Miller.[1] He attempted to find a viable socialist political economy in an age in which the most obvious alternative to liberal capitalism, the centrally planned economies of state socialism, seemed to have been comprehensively shown to be non-viable, and in which the ascendancy of increasingly radical forms of economic liberalism appeared to be unchallengeable. The theoretical arguments of Hayek, which establish what appears to be a conclusive epistemological case against the wholesale replacement of markets with central planning, are particularly important in this regard. In a nutshell these arguments amount to the claim that the knowledge needed by central planners to determine the productive and distributive activities of a large-scale complex economy is not available to them even in principle, for it concerns the needs and desires of millions of individuals, which only those individuals can know at any given moment.[2]

Miller accordingly tried to take account of the historical and theoretical reasons for saying that a market system of some kind is indispensable to a viable economy, while showing, for the benefit of socialists, that such a system could meet the requirements of distributive justice and the need to avoid, or at least minimise, the evils for long held to be central to capitalism, namely exploitation and alienation. Against economic liberals it had to be shown that an alternative market system to that of capitalism could be coherently described which could possess much of the dynamism of capitalism.

Although ecologism has an interest in these matters from the point of view of human well-being alone, deriving from its basic concern with the conditions of human flourishing, it also needs to be satisfied that any alternative to capitalist markets is better able than capitalism to satisfy the requirements of ecological justice. This is an issue which, not surprisingly, Miller did not significantly address, given the aims just outlined.[3] Can we supplement his discussion in this area, or must we conclude that on this matter market socialism and market capitalism are indistinguishable?

The 'pure' model of market socialism with which Miller is concerned is outlined as follows:

> all productive enterprises are constituted as workers' co-operatives, leasing their capital from an outside investment agency. Each enterprise makes its own decisions about products, methods of production, prices, etc., and competes for custom in the market. Net profits form a pool out of which incomes are paid. Each enterprise is democratically controlled by those who work for it, and among the issues they must decide is how to distribute income within the co-operative.[4]

The reasons for evaluating this version of market socialism positively from the point of view of ecologism are that it does attempt to build in procedures of economic decision-making which *prima facie* attend to the autonomy needs of participant workers, and allows the wider society a stake in the productive activities of its members via the control exercised by a State Investment Agency (or Agencies) of the allocation of resources for investment.[5] Further, Miller's case for saying that market socialism reduces the possibility of exploitation and alienation as compared with capitalist markets has some plausibility, and a concern for human flourishing has to involve a concern to reduce these phenomena.[6]

However, as we noted in Chapter 9, a case can be made from the point of view of ecologism for saying that the avoidance of alienation is not necessarily the most important consideration in the working of economic systems. If the preservation of the conditions of existence and flourishing of the non-human requires the toleration of human economic practices which are to some degree alienating, then that is arguably a price that ecological justice rightly exacts. Much will depend on the exact circumstances in a given instance as to whether this view is justifiable from the point of view of ecologism.

The reasons for concern about whether or not market socialism is better able than capitalism to fulfil the requirements of ecological justice stem, obviously, from its market character. Worker co-operatives are expected to compete and to make a profit. They are not to be rescued from bankruptcy if their productive activities fail in the market-place, for the exercise of consumer choice (exit, again) as an essential method of determining allocation of productive resources is retained from market capitalism and is what enables the State Investment Agency to possess the requisite knowledge of where society would prefer investment to go.[7]

Will worker co-operatives be willing to alter or amend their productive practices in order to meet the requirements of ecological justice, rather than to meet the purely anthropocentric aim of keeping human beings in employment? There is no *a priori* reason to suppose so, for such justice requirements are extrinsic to productive activities and thus can be ignored without direct jeopardy to the well-being of a given group of workers. The laying-off of productive workers, by contrast, is an inherent possibility in any kind of market economic system we choose to contemplate, and thus must be entertained as a possibility by those operating the system (such considerations also apply to the working of economic democracy within a capitalist system, of course).

Thus, there seems to be no reason drawn from the very structure of the market socialist system, at least as described by Miller, to suppose that the productive units themselves will have an inbuilt tendency to attend to the requirements of ecological justice. They may be more easily made to do so by the coercive power of the state than is possible under capitalism. But the proposal to use the State Investment Agency for this purpose runs up against bureaucratic politics. A guardianship body within a market socialist economic system may well find that it has a difficult job of ensuring that a powerful agency devoted to seeking profit-making opportunities within a market system pays much more than lip-service to the requirements of ecological justice. These may be no greater difficulties than are encountered in capitalism, where powerful private corporations also offer such resistance, but arguably they may be no less.

Market socialism, then, does not appear from the point of view of ecologism to be the panacea for the environmental ills perpetrated by the capitalist and state socialist systems. It is the promised improvement in human well-being which makes it attractive to ecologism, on the assumption that it does amount to a system with the ability to deliver the goods necessary for human well-being with a much reduced tendency to alienation and exploitation. But its ability to offer a significantly superior performance in the area of ecological justice is clearly open to question.

Global Ecology

The most radical critique of the possibility of sustainable development, whether as articulated by environmental (neo-classical) or ecological economics, has been put forward by supporters of the view which has been dubbed 'global ecology', after the collection of articles of that name edited by Sachs.[8] This view represents a relaunch of the environmental critiques prominent in the 1960s and 1970s which, according to the view, have been subverted and rendered harmless to the forces of capitalist-motored economic growth by the arrival of the rhetoric of sutainable development subsequent to the Brundtland report.

The critique of sustainable development embodied in this view may be expressed in a number of negative and positive positions. The following summary is based on the ideas propounded in the Sachs volume.[9]

(1) On the negative side, global ecology first supports the abandonment of development as the aim of human economic systems, primarily because of its renewed emphasis on the limits to growth inherent in a finite planet.[10] The idea of sustainable development is analysed as a rhetorical device to permit the hegemony of global capital and enhance the power of 'global' politicians, bureaucrats and 'experts', all of whom are drawn from the 'advanced' world of capitalism. Even the distinction between growth and development made by those who are critical of Western notions of economic growth, such as Herman Daly, is rejected as an impossible one.[11] The obverse of this anti-growth position is the emphasis on the need to foster modes of political economy which are simpler, more sustainable in a fully ecological sense, less or anti-consumerist and focused on use-value rather than exchange value.

(2) The processes of international environmental regime-building associated with such events as the Stockholm and Rio conferences are analysed as at best useless and as more likely than not to be positively harmful to environmental concerns. Their sole outcome has been to ensconce an international, capitalist-

based, Western-dominated elite in a position of managerialist power, offering panaceas which involve the technology and scientific knowledge of Western experts and corporations and which do nothing significant to deal with the 'global' environmental issues of global warming and the rest which are their ostensible concern.[12] Their main outcome has been to draw some, mainly Western-based, Non-Governmental Organisations (NGOs) into the embrace of Western states and corporations so as to give a superficial respectability to the proceedings.[13]

(3) Global ecology claims that the idea of 'development' embodied in 'sustainable development' remains that based on the view of a single economic path into the future, namely the one already taken by Western liberal capitalist societies, which the 'underdeveloped' regions of the world are to be encouraged to follow, albeit now in some mysteriously 'sustainable' manner. In fact, it will be the West which keeps the 'development' side and the Third World which will supply the 'sustainability' in the form of providing the sinks for Western wastes and the curtailment of its own economic growth in order to compensate for the environmentally harmful practices of the west.[14]

(4) The onesidedness of 'sustainable development' rhetoric is disguised by its employment of the metaphor of 'one Earth' and of its continual reference to an undifferentiated 'humanity' said to be faced with a single set of problems which all human beings have to co-operate together to solve. The fact that the problems have been overwhelmingly caused by the West and that only if it changes its ways quickly and radically is there any hope of dealing with them is conveniently hidden by this rhetoric. The case for development is even bolstered by the argument that the main threat of environmental destruction now comes from the poverty of the Third World, and that the poor of those countries need to move rapidly further along the Western-style growth path in order to be able to deal properly with 'their' environmental problems.

(5) Global ecology argues that in the past the nature of Western 'development assistance' has usually been positively harmful, and certainly inappropriate, to the needs of the ordinary citizens of the recipient countries. Similarly, the kinds of technological fixes tied into such post-Rio developments as the World Bank's Global Environmental Facility represent the imposition of inappropriate solutions concocted at the global,

state-centred level of international managerialism, by scientific and other Western-trained 'experts'. The net beneficiaries of all this remain the usual ones – Western Multi-National Corporations (MNCs) and Third World elites.[15]

(6) A specific target of criticism is the practice, inherent in the process of capital accumulation as traditionally conceived in capitalist systems, of expropriation and conversion into state or private property of the remaining commons, especially those which remain in the Third World countries.[16] The positive view of the role of commons is set out below.

(7) The critique of the pernicious effect of Western science and expertise is extended to encompass even the science of ecology. In its most recent varieties, it is said to have been unable to give any firm guidance on the issue of what human beings can do to nature without disrupting and destroying natural systems. Scientific ecology is now claimed to have abandoned its earlier certainties about the existence of stable patterns and climaxes in the natural world and to be portraying the latter as devoid of any inherent balance. Where there is no equilibrium or balance there can be no disequilibrium or imbalance. Thus, the only scientific discipline that looked capable of offering some reasoned check to the harmful expansivism of science-based technology and economics in general has turned into a paper tiger.[17]

(8) As is apparent from the earlier points, global ecology rejects or is profoundly suspicious of universalist discourse, whether scientific or moral-political, of the kind used to underpin the practices of international debate on environmental and related matters, and to formulate the 'solutions' to 'global' problems in international forums. It thus takes up again some of the earlier environmentalist cudgels against the Enlightenment inheritance of universalist conceptions of reason. It counterposes to this Enlightenment version of reason the ideas of immanence, diversity – of languages, forms of knowledge and cultures – and emphasises the impossibility of blending these together into a single universal discourse.[18]

The positive conception of an ecologically sustainable form of political economy, which takes proper account of the moral considerability of the non-human, has inevitably been alluded to in the above critique. It involves a version of the communitarian outlook which we have already encountered in previous chapters, and may be summarised as follows.

The ideal form of human socio-economic organisation for global ecology is closely modelled on the traditional, commune-sized and largely agrarian form still to be found in parts of India.[19] Such communes are viewed as having close interconnections with the specific natural environments within which they are located. These environments provide the resources, of course, to sustain the lives of communities, but also have a sizeable impact on the development of the cultures of the communities which interact with them. The members of such communities develop traditions, customs and other practices which enable the community to preserve the ecosystem within which they exist and thereby to preserve the biodiversity which is bound up with it.

The knowledge which such communities possess is not, of course, scientific, theoretical knowledge of the kind which is so celebrated in the West, but is a practice-based knowledge, developed over long periods, and highly attuned to the needs of the complex ecosystem which the people inhabit. Such knowledge is a communal possession, embedded in the culture of the specific commune. Hence, there is a close connection between the preservation of non-human diversity, or biodiversity, and human cultural diversity. Contrariwise, the flattening-out of human cultural diversity and the reduction in biodiversity are also interlinked.

The impact of specific ecosystems on specific human cultures involves the interlinking of practical requirements, such as the need for periods when the land is left fallow, and religious practices and mythic constructs which interweave the practical with the symbolic, aesthetic and social – as in certain festivals and ceremonies. The specific ecosystems within which such traditions develop and have their point furnish the natural patterns, symbols and experiences which give the traditions their concrete form as well as their practical purpose. Western observers, imbued with rationalistic, theory-laden preconceptions, will tend to view such events as outmoded primitivism, picturesque but doomed to be swept away, or perhaps preserved for the benefit of tourism, once 'progress' and 'development' have been set in train.

The specific mode of political economy which is inherent in the traditional commune is the commons. The resources upon which the commune depends are viewed as communally owned and access to them is regulated by the commune itself, via a series of traditional rules and practices accepted by all the members. As we noted in Chapter 6, this is emphatically not the same as the commons famously analysed by Garret Hardin, which is in reality an 'open access' regime, in which access to the resources is completely uncontrolled. Commons properly so called are well adapted to the preservation and biodiversity of the ecosystem involved. Thus, they do not need the imposition of a state-based mode of coercion in order to prevent self-interested individuals from being trapped in a prisoner's dilemma, nor do they need to be enclosed and constituted as private property so as to be 'looked after properly' and 'developed rationally' in the manner beloved by liberal capitalism. Both of these favourite remedies of the West lead instead to the ecologically harmful and culturally catastrophic imposition of monocultures, destroying species and depriving people of their history, culture and identity.

The aim of global ecology, therefore, is to seek to defend the commons-based communities which still exist, albeit in dwindling numbers, against the evils of enclosure and 'development'. Since the whole world is now divided up into states, this means in practical political terms seeking to protect and enhance as far as possible the autonomy of communes from central government, dominated by Western-oriented elites, and international organisations. It also means resisting the favourite Western state-directed solution for the protection of biodiversity, namely the creation of national parks and conservation areas from which the traditional peoples are excluded and/or with respect to which they are prevented from practising their traditional techniques of environmental management.

This is not to say that all such communes are perfect enclaves within a wicked world. However, they are held to be a sounder basis for the preservation of the non-human world and of human well-being than are the consumerist mass societies of the liberal capitalist world. The main proposals put forward by global ecologists for the improvement of traditional communes are: their thorough-going democratisation; their voluntary reorganisation along more explicitly bioregional/ecosystem lines, overcoming the arbitrary frontiers created by state bureaucrats; the development of new forms of intercommunal relationships, democratically developed and aiming to provide communes with mutual aid to fill gaps in each other's natural resource base.[20]

The standard of living attainable in such communes is expected to be one which achieves a reasonable sufficiency for human beings, but which nevertheless will be frugal in comparison with the commodity-glutted West. The gains achievable by communes in the cultural and spiritual richness of life associated with the preservation of a diverse non-human world, humanity of scale and the experience of rootedness, are expected to furnish more than ample compensation for the simplicity of material existence within them.

Let us now consider some of the difficulties with global ecology as outlined above. First, one may properly wonder how accurate are the views expressed by devotees of global ecology concerning the ecologically beneficial nature of traditional agrarian communes. There is certainly evidence that even in precolonial times societies organised on such a basis have played their part in major extinctions and destruction of habitats – consider the extermination of flightless birds by the Maoris in New Zealand, for example. There may be a fallacy lurking in the global ecologists' celebration of the closeness of ecological–cultural interconnections in traditional societies. This is to suppose that because human culture and non-human nature are closely interwoven in such societies all of the non-human will receive care. It is at least possible that such interconnections are often rather partial, with some aspects of the non-human natural world being accorded concern, but others being neglected or attacked, particularly if they have no obvious economic or cultural value to the people involved.

To avoid the charge that global ecology is peddling the myth of the golden age, it would be as well for its supporters to admit the probable deficiencies of all human societies in terms of their ecological soundness. The case for communes and commons is one which need only refer to a marked difference in degree of ecological harmfulness between traditional and modern systems. Differences of degree are often of vital importance.

Secondly, one may well wonder whether all development, even that associated with the Western concept, is to be shunned, especially if one is concerned to promote the conditions for well-being and flourishing of the human species. Traditional communes in agrarian societies such as precolonial India may have been rather more agreeable places than they have become in a modern world in which the basis of their subsistence has been undermined by remote and bungling bureaucracy and/or market-based economic pressures. However, they did face chronic problems of malnutrition, famine, high infant mortality, short life-expectancy, disease, squalor, drudgery and monotony of diet as well as being characterised by such social pathologies as sexism, ignorance, fear and oppressiveness. At any rate, many late twentieth-century inhabitants of these traditional communities do conceive of the latter as prey to such deficiencies. They are also aware of the Western model of development as promising to remedy some or all of these deficiencies, even if at the expense of destroying ancient traditions and despoiling ecosystems.

It looks, therefore, as if an alternative, ecologically sustainable form of community based on commons must be prepared to argue for the necessity for schools, health care and appropriate technology to relieve the tedium and drudgery of traditional communal life. Global ecology aims also to create democratic, participatory and tolerant communities which, while cherishing their own traditions, are open to those of other communities and are not closed in upon themselves. It is hard to see why these desiderata should not be spoken of as development or progress, and thus it seems clear that there has to be room for at least some developmental purposes to be admitted, and ones which are clearly part of the universalist, Western version of the concept too.

In any case, the elites of non-Western countries are clearly committed to 'catching up' with the West – a project understood almost completely in the terms established by liberal capitalism. There may be some residual concern about the loss of distinctive cultural traits in the process, but not, apparently, to the degree necessary to lead such elites to try to put a halt to the 'catching-up' process. Among the non-elite of countries such as India, there may be greater misgivings about the impact of this process on traditional cultural values. However, many of the ordinary people based in traditional communes have already found themselves having to leave the land for the cities in order to try to acquire a wage necessary to purchase what they once were able to grow for themselves, as more and more agricultural land is turned over to lucrative cash crops and monoculture food crops, sold in the national market-place by the new farming elites.

In the cities, of course, country people are relentlessly exposed to Western consumerist, individualistic values and notions of progress, assuming they have not already encountered them in the countryside. In countries such as India and China, where entrepreneurial values and skills have always had a part to play in the life of the society, the new commercially minded groups experience the emphasis of their governments on economic growth and commercial success as a liberation from ancient shibboleths and failures, not as something to be resisted.

The point of these reflections is, obviously, that global ecology's emphasis on the importance of communities and communes, even in those societies where they still may be found, may have arrived too late to be capable of strengthening traditional forms against the undermining tendencies of powerful economic and cultural forces. However, if ecologism is correct in its view that Western models of development do portend ecological disaster, then this may be sufficient to turn what at present appears to be an almost unstoppable force into a passing fashion. In countries such as India, the ideas of Gandhi may eventually win the battle with the ideas of Adam Smith.

A third problem with the global ecologists' diagnosis is the lack of any very convincing prescriptions for what should happen in those societies in which it is virtually impossible any longer to find examples of commons and traditional communes. It is at least intelligible to propose the attempt to preserve existing communes and commons, but how does one re-establish such things when they have gone almost completely – apart from the attempts of environmental or religious groups to create communes within the interstices of Western liberal democracies? Of course, from the point of view of a concern for biodiversity, the global ecologists' strategy at least has the merit that the societies within which traditional communes are still to be found are the very ones which are crucial for maintaining biodiversity. Hence, preserving traditional modes of organisation in these may appear to be both practicable and vital. The 'developed' world has already foregone many of its ecological riches in pursuit of its materialist purposes, so much less importance may attach to trying to recapture the ecologically beneficial mode of political economy in such societies.

However, the problem with this is, of course, that the adverse effects on the world environment associated with global warming, ozone depletion and so on are largely the responsibility of such countries' activities. Also, these countries, with the dynamic capitalist market system at their heart, are inherently expansionist and dominate the world economic system, exploiting the traditional commune-based societies for their raw materials and sink-providing capacities. Clearly, if they remain on their present trajectory there is no real prospect of erecting a great wall of China against their ecologically disastrous activities. They must be greened in some way, and global ecology needs an account of how that can be done effectively and quickly.

A related difficulty with the global ecologists' analysis is that it is silent on the question of how the ideas of the commons and communal organisation may be applied to urban existence, where an increasing proportion of the human population lives, in both the 'developed' and the 'underdeveloped' regions of the world.[21] Of course, the idea of 'commons' regulated by a 'commune' is not totally inapplicable to city life – medieval guilds may provide an example. However, the modern city is one where productive resources are overwhelmingly in private hands and where most people live individualised modes of existence as wage earners.

Further, the attraction of the agrarian commune as a model of economic existence is that it contains a clear conception of the interpenetration of ecology and culture, for on that model specific communes relate to specific ecosystems, which they both manage and are shaped by. However, modern city life is the paradigm of a mode of existence in which human beings are cut off almost entirely from daily interaction with natural ecosystems. It is a strength of the global ecologists' analysis that this explains some of the anomie and cultural uniformity of the city dweller's life, but it is a weakness that it has nothing to say about how the milieu within which a majority of the planet's human inhabitants will shortly be living should be altered in order to develop the kind of cultural/ecological knowledge upon which the future environmental health of the planet depends.

It would seem an implication of the global ecologists' analysis that cities, especially the huge conurbations within which many human beings now live, should ideally be diminished in size, in terms of their populations, their geographical extent, the amount of energy and resources they consume, and the amount of consumer goods and associated waste which they produce. This may involve somehow inducing a voluntary return of people to a countryside organised along commune lines – the example of enforced rustication of people for ideological reasons provided by Cambodia under Pol Pot causes a shudder at this point. Conceivably the aims of global ecology might be met by the penetration of city life by agrarian activities, such as city farms, and the spread of large areas of reclaimed habitat within city limits. Without some sort of an account of what is to be done about cities the global ecologists' analysis is bereft of a strategy for dealing with a massive and accelerating human development which promises to swamp any gains for ecological well-being made in the remaining agrarian parts of the world not yet taken over by city-based agribusiness.

Finally, there are some objections one may have of a more theoretical kind to the analysis provided by global ecology. First, the analysis endorses a view of human culture, encompassing values and conceptions as well as artefacts, which emphasises boundedness and non-translatability, which sits ill with the use of such universalist concepts as democracy, justice and autonomy in its prescriptions for the relations between communes and states. In its recognition that actual examples of communes are far from perfect in many ways – they are subject to forms of oppression and authoritarianism as well as sometimes being properly subject to adverse criticism of the precise nature of their relations to the natural world – global ecologism appears to adopt a transcendent standpoint which its substantive theory of language and culture rules out as impossible.

In its own critique of the relations between liberal capitalism and Third World cultures, the former are depicted as straightforwardly dominating and exploiting the latter in terms of concepts of domination and exploitation which are intended to have cross-cultural applicability. That is, the diagnosis of exploitation is one which both members of the exploiting group and those exploited are expected to recognise as correct, and as correct in the same way. Global ecologists, whatever their native culture, are expected to concur with this diagnosis, which accordingly appears to be culture-transcendent. But if certain universalist concepts are admitted as legitimate it is not clear why others should be rejected. Their universalism cannot itself be an objection.

In any case, there is a worrying reification apparently at work in the global ecologists' account of culture-ecosystem relations. Both specific ecosystems and the specific cultures which they support and with which they interact are depicted as having a fixed boundedness which is inapplicable to either phenomenon. It is of course possible to individuate cultures and ecosystems for certain purposes and within certain rough limits, but one great theme which ecologism has correctly emphasised is the interconnectedness of phenomena, whether human or non-human. Specific ecosystems and human cultures are interconnected across time and at a time. Boundaries are drawn for certain purposes and redrawn when those purposes change.

It is possible that the emphasis on diversity and distinctiveness is not intended by global ecologism to rule out such interconnectedness, and that the aim is simply to redress the balance somewhat in a discussion which has become too 'global' in its orientation. But there is clearly the danger in taking this line that the non-human world will not be treated as a unified whole when in certain respects it is important so to treat it, and that the interconnections between human beings, moral and cultural, will be neglected or downplayed to the detriment of both culture and morality.

Finally, we must consider the rather surprising rejection of scientific ecology as any sort of a guide to how human beings should interact with the natural world so as to further their own well-being without significantly reducing the well-being of non-humans. It may be true that the most recent theorising in the science of ecology has jettisoned any idea of the natural world as exemplifying patterns of stability and equilibrium which give human beings a benchmark against which to measure their impact on the non-human system. It does not follow from this that nothing human beings do can adversely affect the non-human world, and that there are no ecological reasons for objecting to at least some human activity carried out in the name of development. After all, human activity has demonstrably brought about the extinction of some species, has altered habitats to the detriment of both humans and non-humans, and has even altered large parts of the purely physical environment of the planet, such as the ozone layer, weather systems, wetlands and water courses.

In deciding whether or not these are causes for congratulation or regret it may no longer be possible to refer the changes to some theoretically ascertainable situation of equilibrium which non-human nature would have attained in the absence of human activity. But it is clear that key ecological concepts, such as those of ecosystem, negative and positive feedback, keystone species, adaptedness, niches and so forth will need to be employed in order to understand what has happened, what the implications for the future are, and what the effects are likely to be on the well-being and even existence of life-forms. Consider, to take only one example, the crucial importance for the assessment of extinction rates, of the formula for determining numbers of species in island biogeography. Hence, the scientific theorising embodied in ecology is something which cannot reasonably be rejected in favour of the practical knowledge inherent in the traditions of agrarian communes, valuable though that undoubtedly often is.

Now that we have reviewed the case made by global ecology and considered some of the difficulties which the view faces, we need to decide how much of it is defensible. The following points are of value in global ecology.

It is reasonable to be suspicious of the concept of 'globalism' as a conduit for Western, capitalist-based, technocratic managerialism which has a tendency, via international organisations and regimes, to impose pseudo- or non-solutions to an environmental crisis which, up to the late twentieth century, has largely been of Western making. The main beneficiaries of such 'solutions' are probably disproportionately Western-based corporations and parts of Western populations (understanding 'benefit' in traditional economic terms). It is correct to insist that the crisis is one in which the culprit is not 'humanity' and in which not all members of 'humanity' suffer equally, even with respect to such 'global' phenomena as climate change.

The obverse of this position also has a great deal to be said for it, namely that genuine solutions to environmental problems are usually best found at the local level, using local knowledge of specific ecosystems and engaging the interest and concern of people directly involved with them, rather than centrally imposed solutions, based on abstract theory, which involve the attempt to separate local people from the ecosystems in question.

The political implications of this are also valuable, namely the case which is made for decentralisation of decision-making and the greater autonomy of localities from the central state, particularly when coupled with the analysis of the self-regulatory nature of the commons as a mode of economic organisation in an agrarian milieu. The preservation of existing commons and traditional communes is certainly an important consideration in the attempt to see how human and non-human interests may be brought into some kind of harmony.

There is an interesting parallel here to the argument we noted at the start of this chapter which was put forward by Hayek in favour of market-based decision-making and against central planning of the economy. He argued that the knowledge of preferences upon which allocative decisons would have to be made can exist only in the minds of the individuals whose preferences are involved. Such knowledge cannot be gathered into one place to form the basis of a grandiose economic plan. Similarly, global ecology reminds us of the enormous multiplicity of ecosystems which underly scientific ecology's generalisations. This implies that the knowledge of the interactions which make up the actual biosphere is to be found only in the minds of people who are interacting with those systems at the micro level, and thus it is to them, rather than to a central body of scientifically informed experts, that we should look for solutions to serious environmental problems.

Such knowledge is inevitably partial, but has the great advantage that it is derived by methods of direct interaction which enable its possessors to track the specific changes which are ceaseless in any such system, and to accumulate an inductive understanding of what is likely to harm or benefit the local system. For this to work properly, of course, the people operating at this level need an incentive to engage in the effort to discover how their local ecosystem works, which is a further reason for emphasising the importance of the commons as a mode of economic organisation.

This implies in turn that the proper role of the central state with respect to the implementation of this environmental knowledge, embedded in specific cultural formations, is to act as a facilitator rather than as an imposer of solutions. One valuable service it can perform is to protect the localities against those forces which threaten to disrupt the cultural formations involved and to provide environmental information not available to the specific localities as well as fostering their co-operative relations with neighbouring communes.

However, these claims leave intact the case for the creation of a centralised 'guardianship' institution, to monitor changes in the structure and practices of communes in societies where they are in existence. Where they do not exist, and cannot easily be revived, the role of the guardianship institution remains vital, and it must perforce be one which is largely informed by the theoretical analyses of scientific ecology and allied disciplines.

The strength of the case for the commons as (at least potentially) an ecologically sound mode of political economy does support the project of searching for ways to protect commons and communes where they exist in capitalist-dominated societies and for seeking ways to extend their activities. But this, as we noted above, is going to be an extremely difficult task in a West oriented so wholeheartedly to individualism, markets, consumerism and technological innovation.

The general case against the rhetoric of development, sustainable or otherwise, is a valuable reminder of how radical modern environmental critiques are. For development and progress are the great Enlightenment aims, the hallmark of modernity, and in challenging these notions environmental critiques penetrate to the heart of the issue of what humanity should be aiming to achieve on this planet – chiefly by making us suspicious of any rhetoric which postulates that 'humanity' as such has any single aim at all. The renewed emphasis on limits to 'growth' and the justifiable scepticism about whether Western models of growth have been helpful to the well-being of Third World people and their environments are other valuable features of global ecology.

However, it is harder to accept that all that goes under the name of 'development' is equally harmful. Some changes are clearly necessary for the improvement of human flourishing in the Third World – and in parts of the liberal capitalist world too. These changes may be detachable from the model of development which supposes that the path followed by the West is the one true path which all must follow, but there will inevitably be some overlap with it. Health care, education, improved nutrition, basic infrastructure – transport and communications, sewerage and clean water supplies – all are essential to improving the well-being of human beings in even the most ecologically exemplary communes.

Arguably so too are some systems of social welfare to help in cases of old age and illness. The commune may well be able to provide welfare assistance on the basis of mutual aid and solidarity, but that will depend on how close to subsistence level any particular commune is. Communes may expect the individual family to be the prime provider of such welfare, in which case if there is poor health care within them, with high levels of infant mortality, we are faced with the phenomenon of people having large families in order to have someone to depend on in old age, one of the prime factors driving population increase in Third World countries.

It is hard to see how these elements of development are achievable without the generation of economic surpluses. These are needed to provide the capital accumulation necessary to finance these developments. The latter require not simply material capital for construction, but the human capital of medical, engineering, educational amd financial skills and associated institutions. This suggests that a purely subsistence form of commune-based economy has to be rejected. But such rejection in turn implies a commitment to a certain amount of economic growth.[22] The problem then is how to achieve this in an amount sufficient to finance the humanly necessary growth without putting into play forces, such as the creation of products for exchange value, the development of markets, specialisation and consumer products, which lead on willy-nilly to the full-fledged Western model.

It is here that a distinction between a purely quantitative concept of growth and a qualitative concept of development of the kind supported by Daly has a point, and it is hard to see how a defensible ecological position can be developed without it, even if in practice it is hard to offer a clear and watertight specification of what it involves. It is, however, a valuable suggestion one might take from the localism inherent in global ecology that there should not be a single overarching concept of development, but that the content of the concept should be determined largely by localities on the basis of their particular cultural formations.

Part of what is likely to be involved in the qualitative concept of development produced in traditional communes is a concept of praxis which downplays the idea of personal autonomy and self-development present in the Western ideal of progress and which instead emphasises the satisfaction to be gained from working within a community, jointly managed with one's neighbours, in harmony with nature. However, it will be recollected that one respect in which global ecology seeks to modify many traditional communes is in the direction of greater democracy and participation, which implies that the ideal of personal autonomy, although downplayed in relation to the ideal of the autonomy of the commune, also has a place.

The upshot of all this is to view global ecology as offering some salutary counters to a concept of sustainable development which is in danger of simply acting as window-dressing for the Western growth mania which has brought us to our present predicament in the first place. Its emphasis on decentralisation, diversity of eco-systems and culture and of their interplay, and on the subordinate role to be played by states; its scepticism of Western technocracy and crypto-imperialism; its celebration and analysis of the ecological wisdom present in many traditional communes and its emphasis on the importance of distinguishing real commons from open access regimes and on the vital necessity to prevent the remaining ones from suffering expropriation in the name of a misconceived concept of economic development, are all valuable.

But it is clear that some of the enlightenment inheritance needs to be retained, even to make sense of some of global ecology's own prescriptions. The state remains important as a facilitator of commons and as a key regulator, with other states, of the open access systems of the planet. Some ecological phenomena cannot be dealt with by localised communes, but need some form of international co-operation. It may be misleading to refer to these as 'global', with the implication that they are not primarily the result of the activities of the Western liberal capitalist societies, but they nevertheless cannot be dealt with on a localised basis. Also the prospects for the creation of communes based on commons in the world of liberal capitalism do not seem very promising, so that the state has to adopt a much more active role in regulating and controlling environmentally harmful activities there.

The aim of ecologism with respect to the liberal capitalist world has to remain that of developing and gaining acceptance for a distinction between growth and development and of seeking to clarify the concept of sustainability in a way which does not simply allow the present show to carry on as before. In this regard, as we concluded at the end of the previous chapter, the concept of sustainability promulgated by ecological economics appears to be the most promising one. Its emphasis on the need to make economic decisions in a manner which involves democratic discussion and debate at least complements the democratic, participatory emphasis of global ecologism.

With respect to the Third World, some elements of growth, and the scientific and technological underpinning which that requires, remain vital to human well-being. Until life-expectancy, child mortality, morbidity and illiteracy rates show dramatic improvement it is going to be very hard plausibly to maintain, even to their inhabitants, that traditional agrarian communes, ecologically estimable as they may be in other respects, are adequate to sustain human flourishing.

Although global ecologism expresses understandable scepticism about science and technology, and emphasises the merits of traditional forms of knowledge, it is clear that scientific knowledge, especially in the form of scientific ecology, remains of vital importance to an environmentally sound future.

What one might conclude is that global ecology aims to produce a greater evening-up in the conditions of the human inhabitants of the planet not by hastening the 'underdeveloped' parts of the world along the trail blazed by capitalist liberalism, but rather by moving both the West and the Third World more towards each other. The

Third World should be developed without following the Western road, and the valuable aspects of the Third World societies – communalism, commons, closeness to the non-human world, greater humility in the face of nature – should be protected and developed within the West.

However, the upshot of this chapter is to leave ecologism with no clear alternative to seeking the transformation of liberal capitalist political economy. Since capitalism undoubtedly exists, is growing in scope and is attracting the allegiance of billions of new adherents, realism, if nothing else, requires that the direction in which ecologism presently must look is towards the extensive and rigorous scrutiny of ways in which to force ecological and environmental justice on to capitalist enterprises.

This is its distinctive position, rather than any specific proposals for a radically new economic system (although, as we have seen, it has reasons for supporting at least some of these). Clearly, this is a large and complex task, but it is of considerable help to ecologism that it knows what it wants to do in the sphere of the human economy, and why it wants to do it. If we are clear about that then the much vaunted human ingenuity may be turned in an environmentally benevolent direction and show us how to do it.

Notes

1. David Miller, *Market, State and Community* (1989).
2. F. Hayek, *Individualism and Economic Order* (1949), pp. 77–8.
3. However, on p. 324 he suggests that a commitment to socialism imposes no specific view concerning environmental protection.
4. Miller 1989, p. 10.
5. Miller 1989 discusses on pp. 310–12 the desirability of more than one agency, and the need to keep them at 'arm's length from the political centre'.
6. See Miller 1989, chapters 6, 7 and 8.
7. See chapter 5 of Miller's 1989 book for a qualified defence of the role of consumer sovereignty in market socialism.
8. Wolfgang Sachs (ed.), *Global Ecology: A New Arena of Political Conflict* (1993).
9. It should be borne in mind that, for the purposes of discussion, the following summary inevitably, and perhaps misleadingly, creates the impression of a unitary point of view held by what are in fact diverse individuals. The ideas outlined are given a favourable reception by

Matthew Paterson in 'Green politics', in S. Burchill and A. Linklater (eds), *Theories of International Relations* (1996), pp. 263–6.

10. Wolfgang Sachs, 'Global ecology and the shadow of "Development" ', in Sachs 1993, p. 6.
11. Sachs in Sachs 1993, p. 9.
12. Nicholas Hildyard, 'Foxes in charge of the chickens', in Sachs 1993, pp. 28–35.
13. Matthias Finger, 'Politics of the UNCED process', in Sachs 1993, p. 46.
14. See Wolfgang Sachs' Introduction to *Global Ecology*, p. xvi.
15. On this and the previous point see Vandana Shiva, 'The greening of the global reach', in Sachs 1993, pp. 149–56.
16. The case of India is discussed in Smitu Kothari and Pramod Parajuli, 'No nature without social justice: a plea for cultural and ecological pluralism in India', in Sachs 1993, p. 226.
17. Donald Worster, 'The shaky ground of sustainability', in Sachs 1993, pp. 136–40.
18. Larry Lohmann, 'Resisting green globalism', in Sachs 1993, pp. 157–67.
19. See, for example, Frédérique Apfell and Purna Chandra Mishra, 'Sacred groves – regenerating the body, the land, the community', in Sachs 1993, pp. 187–206.
20. See, for example, Kothari and Parajuli in Sachs 1993, pp. 233–4.
21. See the report on the UN Habitat II summit in *The Guardian*, on 5 June 1996, where John Vidal reports that by 2025 the number of people living an urban existence is expected to reach 5 billion – double the 1990 figure.
22. Paul Ekins, 'Making development sustainable', in Sachs 1993, pp. 96–7, emphasises this point, arguing that in the 'short- and medium-term at least' there is the need in the South for 'balanced, sustainable growth through a twin focus on environmental regeneration and careful industrialisation using the most environmentally advanced technologies'. The latter implies 'considerable technology transfer from North to South'. However, this view is somewhat at odds with the general tenor of other contributors' views.

12

Conclusion

THE PHILOSOPHER AND NOVELIST Iris Murdoch once suggested that the main aim of philosophy is to find the appropriate context in which to state the obvious.[1] This illuminating comment creates an immediate link between art and philosophy. For both of these activities in which human beings have always been engaged are concerted attempts to attain a proper grasp of what, in one sense, we are all perfectly aware of. A 'proper' grasp here means a clear view of the implications and importance of some phenomenon. In engaging in art and philosophy it is phenomena central to human life upon which we focus. Such phenomena as our own mortality, the limitations of our sympathies, the possibility and actuality of love – and hate – in human relations and so forth are all cases in point. Their existence is news to no-one. But what art and philosophy do, the one by means of discursive argument (but also by literary and rhetorical devices) and the other by harnessing the power of aesthetic qualities, is to focus our minds on these phenomena and bring us to appreciate their full force and importance in human life.[2]

Particular philosophical doctrines in the traditional disciplines of the subject are all attempts to do this. The historically dominant ideologies of the West may all be seen as prolonged meditations on the importance for human life of certain basic facts about us, facts of which we are all perfectly well aware. The different ideologies focus on different such facts as their prime concern, and seek to place the facts made central by other ideologies in the appropriate subordinate position.

Thus, for liberalism the basic fact about us is that we are rational creatures, and the life necessary for such a creature involves rationally defensible social arrangements and the respect for the autonomy which rationality both makes possible and requires. For the conservative, the basic fact is that we are culture-creators, deriving our sense of who we are from our embedding in particular structures of custom and tradition which give us our values and the sense of the possibilities for our lives. For the socialist, the basic fact is that we are social creatures, requiring close interrelations with each other in ways which support our sense of worth and mutual concern, so that social relations which set us apart in relations of hostility and competition damage something central to our humanity.

The exponents of these ideologies do not characteristically deny the facts pointed to by their rivals. Liberals, conservatives and socialists all agree that we are social creatures who create cultures and possess rationality. They disagree over the relative importance of such 'obvious' facts about us, what the terms should be held to mean precisely and what their implications are. But a place is found within each of the theories and their various subvarieties for each of these basic facts about us.

Ecologism is no different. It maintains that the basic fact about us which should be given pride of place is that we are natural creatures. That is, we are a species of animal inhabiting a rich and complex biological context. This fact has either been ignored or consciously downplayed by other political ideologies. Our possession of rationality and the intensity and elaboration of our cultural activities in particular have been held to make us special, different, in some sense set apart from the rest of nature. Ecologism insists that this is not so. We are a species of animal with a highly distinctive set of capacities, but once we begin to view ourselves in the appropriate context, on the basis of our awareness of the rest of the natural world, we can see important kinds of continuity between ourselves and that world.

Facing this fact directly ought to make us more concerned than we have been hitherto about how that world is faring. We ought, too, to become immediately more sensitive to the moral claims which our fellow-creatures have upon us. We ought to become aware of how our well-being and fate are intertwined with those of other living beings with which both recent Darwinian biology and ancient tribal lore (positing ancestors drawn from the non-human world of nature) claim we share a common descent. Ecologism has, therefore, been led, under the prompting initially of a largely prudential concern about the effects of human activity on the planet, to focus upon a basic fact about us to which other ideologies and philosophical positions are held to have failed to play proper attention.

Ecologism, therefore, does not deny that we are rational (and thus moral), social and culture-creating creatures. It holds that these ingredients in our nature are all conditioned by the fact that we are natural creatures. This makes an enormous difference to our understanding of these aspects of human nature. We begin fully to understand the importance of our embodiment as living entities in a network of relationships which we have not chosen and which form the indispensable backdrop to all our activities. We understand our rationality in terms of our needs as a species to make a living in a specific natural environment, and understand how that rationality may get us into trouble when we use it to transform the world in such a way that our evolutionary history starts to become detached from aspects of our way of life. The rationality traps called prisoners' dilemmas start to manifest themselves as we outrun the existing bounds to our productive activities, for example, and come face to face with the misnamed 'tragedy of the commons'.

Our cultural activities become more readily grasped as requiring the context of non-human nature from which we draw ideas, inspiration, metaphors, identities, motifs and upon which is focused the vital, culturally charged, sense of love towards a particular place and landscape. It starts to become more readily apparent that a world made over entirely to become a giant artefact, even if it were attainable, threatens the well-springs of our cultural life and development.

Our sociality becomes similarly nuanced. We begin to see the importance of the non-human natural world as a vehicle for our interpersonal relationships and to see the connections between our own sociality and that of other social creatures. We may thereby start to understand the specific form that sociality takes, and why and where it breaks down. We note the importance, in another of those obvious but not fully noticed facts about us, of the social connections which we seek to establish with the non-human, in our gardening, ownership of pets and enjoyment of domestic animals. For other ideologies, such phenomena are largely meaningless foibles. For ecologism, they represent a glimpse of something fundamental to our self-understanding which we have not fully grasped.

This book has attempted to tease out the implications for the conduct and structure of human life of making the basic fact of our being natural creatures central to our self-understanding. Some will find this a demeaning and intolerable thought. Ecologism counsels such people to attend more fully to what the natural world is like before they conclude that our membership of it is something about which we should keep quiet. Love of the natural world is what makes ecologism an attractive system of thought for many. As with all manifestations of love, the integrity of the feeling requires that we be clear-sighted about the loved one's faults and deficiencies, and be clear, too, about the distinction between love and blind worship. These are intellectual requirements and thus it might be held that they have no place in an emotion. But to think that is to fail to see that for a rational creature such as a human being, feelings are structured by beliefs and can be fostered or obliterated by a change in beliefs.

This book, as is true of all works of philosophy, has been directed towards establishing the intellectual case and implications of certain beliefs about facts and values. But this should not disguise the crucial importance of emotion, and an especially powerful emotion too, namely love, in establishing the claim of the body of doctrine upon our allegiance. It is clear that the intellectual arguments cut no ice unless the feeling is present, or capable of being aroused and held firmly in the centre of one's vision. The intellectual arguments are always fallible, open to objection, never conclusive. But ecologism rests on the belief, perhaps an act of faith or a hunch, that human beings, as natural creatures, are capable of the kind of love of their rich, teeming, beautiful world which leads them to see their responsibility towards it. This in turn should lead them to temper their entirely defensible efforts to secure the good life for themselves and their intimates with a sense of what is needed to keep alive and well the object of their wider love.

If ecologism is wrong about the existence or power of this emotion among human beings, then only self-interest remains to protect the non-human world from our depredations. This may be enough to afford that world the protection it needs, but it is not unreasonable to suppose that it will not be, especially if our undoubted ingenuity enables us to replace natural services with ones of our own devising.

Ideologies sometimes appeal only to the converted. What I have tried to do in this book is to offer arguments which those who are not yet supporters of ecologism will find to sustain a reasonable case, indeed, I hope, a compelling one, for rethinking the basis of our moral theories in such a way as to attribute to the non-human a moral status with which we must reckon in all we do. Even if they are not convinced it ought at least to be obvious that it is not confusion, sentimentality or evil intent which gives rise to this belief.

Rather, once it has been articulated its truth will, I believe, be thought to be obvious.

Notes

1. In conversation with Bryan Magee as part of a series on the great philosophers broadcast by the BBC in the late 1970s.
2. I discuss the implications of these claims further in my 'Art and embodied truth' (1983).

Bibliography

Aitken, Gill (1996), *Extinction*, Lancaster University: Thingmount Working Paper TWP 96–02.

Allaby, Michael (1996), *The Basics of Environmental Science*, London and New York: Routledge.

Archer, Robin (1995), *Economic Democracy: The Politics of Feasible Socialism*, Oxford: Oxford University Press.

Attfield, Robin (1991), *The Ethics of Environmental Concern*, 2nd edn, Athens, Ga., and London: The University of Georgia Press.

Attfield, Robin and Andrew Belsey (1994), *Philosophy and the Natural Environment*, Cambridge: Cambridge University Press.

Barry, Brian (1996), 'Circumstances of justice and future generations', in R. Sikora and B. Barry (eds), *Obligations to Future Generations*, Cambridge: White Horse Press.

Barry, John (1996), 'Sustainability, judgement and citizenship', in B. Doherty and M. de Geus (eds), *Democracy and Green Political Thought*, London: Routledge.

Baxter, Brian (1983), 'Art and embodied truth', *Mind*, XCII, pp. 189–203.

Baxter, Brian (1996), 'Ecocentrism and persons', *Environmental Values*, 5, pp. 205–19.

Baxter, Brian (1999), 'Environmental ethics – values or obligations?', *Environmental Values*, forthcoming.

Beck, Ulrich (1992), *The Risk Society: Towards a New Modernity*, London: Sage.

Beckerman, Wilfred (1974), *In Defence of Economic Growth*, London: Cape.

Beitz, Charles (1979), *Political Theory and International Relations*, Princeton, NJ: Princeton University Press.

Benton, Ted (1991), 'Biology and social science: why the return of the repressed should be given a (cautious) welcome', *Sociology*, 25, pp. 1–29.

Benton, Ted (1993), *Natural Relations: Ecology, Animal Rights and Social Justice*, London: Verso.

Booth, Douglas (1997), 'Preserving old-growth forest ecosystems', *Environmental Values*, 6, pp. 31–48.

Callicott, J. Baird (1989), *In Defense of the Land Ethic: Essays in Environmental Philosophy*, Albany: SUNY Press.

Carter, Alan (1993), 'Towards a green political theory', in A. Dobson and P. Lucardie (eds), *The Politics of Nature*, London and New York: Routledge.

Christoff, Peter (1996), 'Ecological citizenship and ecologically guided democracy', in B. Doherty and M. de Geus (eds), *Democracy and Green Political Thought*, London: Routledge.

Cooper, David (1995), 'Other species and moral reason', in D. Cooper and J. Palmer (eds), *Just Environments*, London: Routledge.

Cooper, David and Joy Palmer (1995), *Just Environments*, London: Routledge.

Daly, Herman and John Cobb (1990), *For the Common Good*, London: Green Print.

Davies, Paul (1983; 1990), *God and the New Physics*, London: Dent; Harmondsworth: Penguin Books.

de Geus, Marius (1996), 'The ecological restructuring of the state', in B. Doherty and M. de Geus (eds), *Democracy and Green Political Theory*, London: Routledge.

Devall, Bill and George Sessions (1985), *Deep Ecology: Living as if Nature Mattered*, Salt Lake City, Utah: Peregrine Smith.

Dickens, Peter (1992), *Society and Nature: Towards a Green Social Theory*, London: Harvester Wheatsheaf.

Dietz, Frank and Jan van der Straaten (1993), 'Economic theories and the necessary integration of ecological insights', in A. Dobson and P. Lucardie (eds), *The Politics of Nature*, London and New York: Routledge.

Dobson, Andrew (1995), *Green Political Thought*, 2nd edn, London and New York: Routledge.

Dobson, Andrew (1996a), 'Democratising green theory', in B. Doherty and M. de Geus (eds), *Democracy and Green Political Thought*, London: Routledge.

Dobson, Andrew (1996b), 'Representative democracy and the environment', in W. Lafferty and J. Meadowcroft (eds), *Democracy and the Environment*, Cheltenham and Brookfield, Vt.: Edward Elgar.

Dobson, Andrew and Paul Lucardie (eds) (1993), *The Politics of Nature*, London: Routledge.

Doherty, Brian and Marius de Geus (eds) (1996), *Democracy and Green Political Thought*, London: Routledge.

Doyle, Timothy and Doug McEachern (1998), *Environment and Politics*,

London: Routledge.
Dryzek, John S. (1990), *Discursive Democracy: Politics, Policy and Political Science*, Cambridge: Cambridge University Press.
Dryzek, John S. (1997), *The Politics of the Earth: Environmental Discourses*, New York: Oxford University Press.
Dworkin, Ronald (1981), 'What is equality? Part 1: Equality of Welfare' and 'Part 2: Equality of resources', *Philosophy and Public Affairs*, 10, pp. 185–246 and 283–345.
Easterbrook, Greg (1996), *A Moment on Earth*, New York and London: Penguin.
Eckersley, Robyn (1989) 'Green politics and the New Class: selfishness or virtue', *Political Studies*, XXXVII, pp. 205–23.
Eckersley, Robyn (1992), *Environmentalism and Political Theory: Towards an Ecocentric Approach*, London: UCL Press.
Elkington, John and Tom Bourke (1989), *The Green Capitalists*, London: Victor Gollancz.
Fishkin, James (1982), *The Limits of Obligation*, New Haven and London: Yale University Press.
Fox, Warwick (1990), *Toward a Transpersonal Ecology: Developing New Foundations for Environmentalism*, Boston, Mass.: Shambala.
Gare, Arran (1995), *Postmodernism and the Environmental Crisis*, London and New York: Routledge.
Goldblatt, David (1996), *Social Theory and the Environment*, Cambridge: Polity Press.
Goodall, Jane (1971), *In the Shadow of Man*, Glasgow: Collins.
Goodin, Robert (1992), *Green Political Theory*, Cambridge: Polity Press.
Goodin, Robert (1996), 'Enfranchising the Earth, and its alternatives', *Political Studies*, 44/5, pp. 835–49.
Goodpaster, Kenneth (1978), 'On being morally considerable', *Journal of Philosophy*, 75, pp. 308–25.
Gould, Carol (1990), *Rethinking Democracy*, Cambridge: Cambridge University Press.
Gray, R. H. (1994), 'Corporate reporting for sustainable development: accounting for sustainability in 2000 AD', *Environmental Values*, 3, pp. 17–45.
Harris, Richard (1996), 'Approaches to conserving valuable wildlife in China', *Environmental Values*, 5, pp. 303–34.
Hayek, Friedrich (1949), *Individualism and Economic Order*, London: Routledge.
Hayek, Friedrich (1960), *The Constitution of Liberty*, London and Henley: Routledge & Kegan Paul.
Hayward, Tim (1995), *Ecological Thought: An Introduction*, Cambridge: Polity.
Hayward, Tim (1996), 'What is green political theory?', in I. Hampshire-

Monk and J. Stanyer (eds), *Contemporary Political Studies*, Belfast: Political Studies Association of the United Kingdom.

Hayward, Tim (1997), 'Anthropocentrism: a misunderstood problem', *Environmental Values*, 6, pp. 49–63.

Holbrook, Daniel (1997), 'The consequentialist side of environmental ethics', *Environmental Values*, 6, pp. 87–96.

Jackson, Tim (1996), *Material Concerns: Pollution, Profit and Quality of Life*, London: Routledge.

Jacobs, Michael (1991), *The Green Economy: Environment, Sustainable Development and the Politics of the Future*, London: Pluto.

Jamieson, Dale (1998), 'Animal liberation is an environmental ethic', *Environmental Values*, 7, pp. 41–57.

Johnson, Lawrence (1993), *A Morally Deep World*, Cambridge: Cambridge University Press.

Kahn, Herman and Julian Simon (1984), *The Resourceful Earth: A Response to Global 2000*, Oxford: Blackwell.

Kant, Immanuel (1948), *Groundwork of the Metaphysic of Morals*, trans. by H. Paton as *The Moral Law*, London: Hutchinson & Co.

Kymlicka, Will (1990), *Contemporary Political Philosophy: An Introduction*, Oxford: Clarendon Press.

Lafferty, William and James Meadowcroft (1996), *Democracy and the Environment: Problems and Prospects*, Cheltenham and Brookfield, Vt.: Edward Elgar.

Lawton, John and Robert May (eds) (1995), *Extinction Rates*, Oxford: Oxford University Press.

Leakey, Richard and Roger Lewin (1996), *The Sixth Extinction*, London: Phoenix.

Lee, Keekok (1993), 'To de-industrialise – is it so irrational?', in A. Dobson and P. Lucardie (eds), *The Politics of Nature*, London and New York: Routledge.

Lovelock, James (1979), *Gaia: A New Look at Life on Earth*, Oxford: Oxford University Press.

Low, Nicholas and Brendan Gleeson (1998), *Justice, Society and Nature*, London: Routledge.

Lynch, Tony and David Wells (1998), 'Non-anthropocentrism? A killing objection', *Environmental Values*, 7/2, pp. 151–63.

MacIntyre, Alasdair (1981), *After Virtue: A Study in Moral Theory*, London: Duckworth.

Mackie, J. L. (1977), *Ethics: Inventing Right and Wrong*, Harmondsworth: Penguin Books.

Martell, Luke (1994), *Ecology and Society: An Introduction*, Cambridge: Polity Press.

Mathews, Freya (1991), *The Ecological Self*, London: Routledge.

Mathews, Freya (ed.) (1996), *Ecology and Democracy*, London: Frank Cass.

Miller, David (1989), *Market, State and Community: Theoretical Foundations of Market Socialism*, Oxford: Clarendon Press.

Munda, Giuseppe (1997),'Environmental economics, ecological economics and the concept of sustainable development', *Environmental Values*, 6, pp. 213–33.

Naess, Arne (1989), *Ecology, Community and Lifestyle*, Cambridge: Cambridge University Press.

Niskanen, William (1971), *Bureaucracy and Representative Government*, Chicago: Aldine-Atherton.

Norton, Bryan (1987), *Why Preserve Natural Variety?*, Princeton, NJ: Princeton University Press.

Norton, Bryan (1991), *Toward Unity among Environmentalists*, Oxford and New York: Oxford University Press.

Nozick, Robert (1974), *Anarchy, State and Utopia*, London: Blackwell.

Okin, Susan Moller (1992), *Women in Western Political Thought*, 2nd edn, Princeton, NJ: Princeton University Press.

O'Neill, John (1993), *Ecology, Policy and Politics: Human Well-being and the Natural World*, London: Routledge.

O'Neill, Onora (1997), 'Environmental values, anthropocentrism and speciesism', *Environmental Values*, 6, pp. 127–42.

Ophuls, William (1977), *Ecology and the Politics of Scarcity*, San Francisco, Calif.: W. H. Freeman.

Parfit, Derek (1984), *Reasons and Persons*, Oxford: Clarendon Press.

Passmore, John (1980), *Man's Responsibility for Nature*, 2nd edn, London: Duckworth.

Paterson, Matthew (1996), 'Green politics', in Scott Burchill and Andrew Linklater (eds), *Theories of International Relations*, London: Macmillan.

Pearce, David and Kerry Turner (1990), *Economics of Natural Resources and the Environment*, New York: Harvester Wheatsheaf.

Pepper, David (1993), *Eco-Socialism: From Deep Ecology to Social Justice*, London: Routledge.

Plumwood, Val (1993), *Feminism and the Mastery of Nature*, London: Routledge.

Popper, Karl (1972), *Objective Knowledge: An Evolutionary Approach*, Oxford: Oxford University Press.

Raup, David (1991), *Extinction: Bad Genes or Bad Luck?*, Oxford: Oxford University Press.

Rawls, John (1972), *A Theory of Justice*, Oxford: Oxford University Press.

Regan, Tom (1983), *The Case for Animal Rights*, Berkeley, Calif.: University of California Press.

Rolston, Holmes (1988), *Environmental Ethics*, Philadelphia: Temple University Press.

Routley, Richard and Val Routley (1979), 'Against the inevitability of human chauvinism', in K. E. Goodpaster and K. M. Sayre (eds), *Ethics and the Problems of the Twenty-first Century*, Notre Dame: University of Notre Dame Press.

Ruse, Michael (1986), *Taking Darwin Seriously: A Naturalistic Approach to Philosophy*, Oxford: Blackwell.

Ruse, Michael (1993), *Evolutionary Naturalism: Selected Essays*, London: Routledge.

Sachs, Wolfgang (ed.) (1993), *Global Ecology: A New Arena of Political Conflict*, London and Atlantic Highlands: Zed Books; Halifax, NS: Fernwood.

Sale, Kirkpatrick (1985), *Dwellers in the Land: The Bioregional Vision*, San Francisco, Calif.: Sierra Club Books.

Sandel, Michael (1982), *Liberalism and the Limits of Justice*, Cambridge: Cambridge University Press.

Saward, Mike (1993), 'Green democracy?', in Andrew Dobson and Paul Lucardie (eds), *The Politics of Nature*, London: Routledge.

Shiva, Vandana (1989), *Staying Alive: Women, Ecology and Development*, London: Zed Books.

Sikora, R. and Brian Barry (eds) (1996), *Obligations to Future Generations*, Cambridge: White Horse Press.

Simon, Julian (1981), *The Ultimate Resource*, Princeton, NJ: Princeton University Press.

Singer, Peter (ed.) (1994), *Ethics*, Oxford: Oxford University Press.

Singer, Peter (1995), *Animal Liberation*, 2nd edn, London: Pimlico.

Sylvan, Richard and David Bennett (1994), *The Greening of Ethics: From Anthropocentrism to Deep-green Theory*, Cambridge: White Horse Press; Tucson, Ariz.: University of Arizona Press.

Taylor, Bob Pepperman (1996), 'Democracy and enviromental ethics', in W. Lafferty and J. Meadowcroft (eds), *Democracy and the Environment*, Cheltenham and Brookfield, Vt.: Edward Elgar.

Taylor, Charles (1979), *Hegel and Modern Society*, Cambridge: Cambridge University Press.

Taylor, Michael (1987), *The Possibility of Cooperation*, Cambridge: Cambridge University Press in collaboration with Maison des Sciences de l'Homme, Paris.

Taylor, Paul (1986), *Respect for Nature: A Theory of Environmental Ethics*, Princeton, NJ: Princeton University Press.

Turner, Frederick (1997), *John Muir: From Scotland to the Sierra*, Edinburgh: Canongate Books.

Turner, R. K., D. W. Pearce and I. Bateman (1994), *Environmental Economics: An Elementary Introduction*, London: Harvester Wheatsheaf.

Van Parijs, P. (ed.) (1992), *Arguing for Basic Income*, London: Verso.

Walzer, Michael (1983), *Spheres of Justice: A Defence of Pluralism and Equality*, Oxford: Robertson.

Weale, Albert (1992), *The New Politics of Pollution*, Manchester: Manchester University Press.

Williams, Bernard (1981), *Moral Luck: Philosophical Papers 1973–1980*, Cambridge: Cambridge University Press.

Williams, Bernard (1985), *Ethics and the Limits of Philosophy*, London: Fontana.

Wilson, Edward (1978), *On Human Nature*, Cambridge, Mass., and London: Harvard University Press.

Wilson, Edward (1980), *Sociobiology* (abridged edn), Cambridge: Belknap Press of Harvard University Press.

Wilson, Edward (1984), *Biophilia*, Cambridge, Mass.: Harvard University Press.

Wilson, Edward (1992), *The Diversity of Life*, London: Allen Lane.

Wilson, Edward (1997), *In Search of Nature*, London: Allen Lane, The Penguin Press.

World Commission on Environment and Development (1987), *Our Common Future*, London: Oxford University Press.

Yearley, Stephen (1992), *The Green Case: A Sociology of Environmental Issues, Arguments and Politics*, London: Routledge.

Index

Accountability, 111, 126
Accountancy, 200
Agencies of change, 168–9
Agriculture, 190, 193–4
Alienation, 166–8, 170, 211
'All affected' principle, 206
'All subjected' principle, 206
Altruism, 38, 40, 139–40
Anarchism, 35, 137, 140–1
Animal liberation, 59–61
Anthropocentrism, 2, 7, 52, 63, 66
Archer, Robin, 205–8
Aristotle, 66
Art, 231
Artefacts, 68
Artificial capital, 198
Atomism, 16–18, 21
Authoritarianism, 125
Autonomy, 122, 124, 127, 155–6, 170–2, 182, 206, 227

Barry, Brian, 91–7
Barry, John, 71n, 117–18n
Basic income schemes, 156
Basic needs, 152–3
Beck, Ulrich, 203
Bennett, David, 44
Benton, Ted, 36
Biodiversity, 2, 19, 28–31, 133, 190, 193, 195
Biophilia, 38, 41–5, 67, 109–10
Bioregions, 138, 217
Biosphere, 52–3, 106
Blue whale, 77
Booth, Douglas, 83
Bourke, Tom, 209n
Brundtland, Report, 197, 213

Callicott, J. Baird, 35
Capitalism, 191, 200–8, 220
Carter, Alan, 136
Central planning *see* planned economy
CFCs (Chlorofluorocarbons), 192, 195
Circumstances of justice, 91–2, 163–4
Cities *see* urbanisation
City farms, 221
Civil society, 137
Common good, 170–2
Commons, 136, 215, 216–17, 225
Communes, 216–22, 225
Communitarianism, 170–6, 215–18
Competitiveness, 204
Complexity of life-forms, 24, 26, 79–80
Consequentialism, 146–8
Conservatism, 232
Consistency arguments, 59–63, 134
Contemporary aliens, 89–90
Contextual view of self, 6, 9, 53, 108–11
Contrarians, 191–6
Cosmic self *see* selves

Cost-benefit analysis, 83
Cultural relativism, 132, 170–1, 222
Culture, 36–7, 44–5, 56–7, 129–30, 138–9, 216, 222, 232–3

Daly, Herman, 123, 227
Darwinian epistemology, 40–1
Death *see* killing
Decentralisation, 224
Deep ecology, 16, 51–2, 110
Degrees
  individuality, 80
  intrinsic value, 74, 76–9
  moral considerability, 52, 79–80
  personhood, 74
Democracy, 111, 120, 121–8, 199, 205–8, 217
Dependants, 182
Development, 218–19, 226–9
Development assistance, 214
'Difference' feminism, 176–8
Distributive justice *see* social justice
Dobson, Andrew, 1, 5, 125–8
Domestic animals, 159, 233
'Dominance' feminism, 176–80
Dryzek, John, 9, 137, 200–1
Dualism, 36, 56
Dutch National Environment Policy Plan, 201–2
Dworkin, Ronald, 145, 149

Easterbrook, Greg, 26
Ecocentrism, 51–2, 60–1
Ecofeminism, 177–8
Ecological economics, 198–200
Ecological justice, 75, 81, 114–16, 125, 148–54, 212
Ecological management, 53
Ecological modernisation, 200–5
Ecological niche, 28–9
Ecologism, 1–11, 22, 27, 30–1, 45, 53, 57, 70, 74, 88–9, 94–5, 124, 127–9, 134, 172, 199, 205, 228–9, 232–5
Ecology, 3, 53, 215, 223, 224–5
Economic democracy, 205–8
Economic growth, 156, 189, 204–5, 213, 226–7
Ecosocialism, 76, 106, 222
Ecosystem, 76, 106, 222
Einsteinian space-time, 18, 20
Ekins, Paul, 230n
Elkington, John, 209n
Enclosure, 217
Endangered Species Act, 113
Enlightenment *see* rationalism
Environmental economics, 197–8
Environmental justice, 75, 114–16, 202
Epigenetic rules, 34, 38–9, 41–2, 45n
Equality of opportunity, 93–4
Equality of persons, 145–6
Ethic of care vs ethic of justice, 180–3
Ethical investment, 203
Ethics of discourse, 130
Exclusivity, 4
'Exit' strategy, 206
Exploitation, 165–6, 211
Exponential growth, 205
Extermination *see* killing
Extinction, 2, 24, 26, 29, 193, 195, 223

Fallibilism *see* scientific explanation
Family, 179–80
Feminism, 176–83
Finitude of Earth *see* limits
Fullness of being, 28–30
Fundamental moral norms, 37
Future generations, 89–97, 126, 173

'Gaia' hypothesis, 19, 60
Gandhi, Moandas, 220
Gardening, 233
Geometrodynamics *see* Einsteinian space-time
Geus, Marius de, 113
Gleeson, Brendan, 75, 110–11, 114–15, 118n, 135, 143n
Global commons *see* commons
Global ecology, 213–29
Global warming, 190, 192, 195, 220
Globalism *see* government
Goldblatt, David, 190
Goodall, Jane, 123
Goodin, Robert, 125, 127
Gould, Carol, 122
Government, 110–11, 120, 128–36, 224
Gray, R. H., 200, 209n

Great apes, 37, 123
Green political theory, 7–8
Green rationalism *see* rationalism
Green romanticism, 9
Greenhouse gases, 189
Guardians, 107, 108–14, 116, 125, 200, 202, 207, 212, 225

Hardin, Garret, 217
Hayek, Friedrich, 210, 224
Hayward, Tim, 7, 8, 52–3
Hegel, G. W. F., 189
Hobbes, Thomas, 106, 108
Holism, 16–18, 31, 77–8
Homeostasis, 17
Hotspots, 133–4
Human arrogance, 8, 17, 53, 55
Human chauvinism, 52, 59
Human dignity, 158
Human ecology, 123–4, 139
Human flourishing, 127
Human nature, 105
Human predicament, 106
Human sociality, 233
Human talents, 154
Hume, David, 91–2

Idealism, 4–5
Ideology, 1–4, 55, 124, 127–8, 169, 231–5
Immanentism, 131–2, 134, 215
Imperialism, 144n
Income streams, 162n
Industrialism, 190–1
Initial acquisition, 156–7
Initial holdings, 93, 95
Interconnectedness, 3, 6, 9, 16, 20, 22, 26–7, 31, 51, 53–4, 75–9, 108–9, 142, 222, 232
Intergenerational justice *see* future generations
International Court of the Environment, 111–12
International distributive justice, 90–1
International environmental regimes, 213
Internet, 135

Island biogeography, 223

Jackson, Tim, 204, 209f
Jamieson, Dale, 81
Johnson, Lawrence, 60
Justice, 90–7, 152

Kant, Immanuel, 65, 92–3
Killing, 23–4, 26, 82, 218
Kin selection, 38, 40–1
Krill, 77
Kymlicka, Will, 145–83

Last human being, 57–8
Liberalism, 109, 134, 171, 232
Libertarianism, 154–61
Liberty, 159–60, 172
Limits, 5, 189
Locke, John, 108, 155
Love, 109–10, 234
Lovelock, James *see* 'Gaia' hypothesis
Low, Nicholas, 75, 110–11, 114–15, 118n, 135, 143n
LULU (Locally undesired land use), 114
Lynch, Tony, 83–9

Market economy, 141, 203, 210–11
Market socialism, 210–13
Martell, Luke, 103, 122
Marxism, 163–70
Materialism *see* idealism
Mathews, Freya, 16–31, 63–6, 69, 76–9
Meaning of life, 20, 23–8
Meta-ethics, 120
Metaphysics, 15–31
Miller, David, 210–13
Minimal state, 137, 180
MNC (Multi-National Corporation), 114, 215
Monism, 18–22, 24
Moral agency, 82
Moral considerability of non-humans, 5, 9, 51–70, 74–89, 148, 151, 178–9, 182, 193, 232
Moral 'toss-up', 78
Moral trade-offs, 74–97, 116
Morally arbitrary features, 149–51

Muir, John, 8, 28
Munda, Giuseppe, 197–9
Murdoch, Iris, 231

Naess, Arne, 16
Nanotechnology, 6, 192
National parks, 217
Nations, 138
Natural capital, 198
Natural/social science distinction, 122–3
Naturalism, 8, 105, 232
Neo-Darwinism, 16
Neutrality of state, 171–5
New class, 169
New Social Movements, 169
NGOs (Non-Governmental Organisations), 214
Norton, Bryan, 6, 8, 53–8
Nozick, Robert, 154–5, 158

'One Earth' metaphor, 214
O'Neill, John, 65–6, 71n, 99n
O'Neill, Onora, 73n
'Open access' regime, 217
Ozone depletion, 190, 192, 195, 220

Particular commitments, 147, 173
Paternalism, 171–2
Pearce, David, 198
Perfectionism, 16, 175–7
Personhood, 121–2
Pets, 233
Philosophy, 231
Piecemeal transformation, 113
Planned economy, 210, 224
Platonic guardians, 121
Plenitude *see* fullness of being
Pluralism, 172–4
Political ecologists, 10
Political ecology, 10–11
Political justice, 143n
Political philosophy, 103–5, 122
Pollution, 189, 192, 194
Popper, Karl, 166
Population growth, 190
Postmodernism, 121, 131, 143n
Pot, Pol, 221
Poverty, 190
Pragmatic democratic politics, 3–4, 54
Praxis, 188–9, 227
Precautionary principle, 201
Preference-satisfaction, 147
Prioritisation rules *see* moral trade-offs
Priority of humans, 83–9
Prisoners' dilemmas, 233
Private property regimes, 136, 141–2, 217
Private/public distinction, 179–80
Profit motive, 203–5
Property-rights, 155
Proxies *see* guardians
Prudence, 29
Public choice school, 136
Public sphere, 137

Rationalism, 8, 9
Rationality, 65, 232–3
Rawls, John, 91–2, 148–51
Raz, Joseph, 173, 176
Reciprocal altruism, 38, 40–1
Recycling, 189
Redistribution, 157
Reflexivity, 123
Regan, Tom, 59
Regulation, 203, 206–7
Religion, 109–10
Renewable energy, 189
Reporting for sustainability, 200
Resource problems, 188–9, 192
Risk society, 203
Rousseau, Jean-Jacques, 122
Routley, Richard, 83
Routley, Val, 83
Ruse, Michael, 38–41, 87

Sachs, Wolfgang, 213
Scientific explanation, 21–2, 127–8, 131, 135, 171
Selves, 18–19, 23–6, 63–6, 76
Self-determination *see* autonomy
Self-interest, 234
Self-maintenance *see* selves
Self-ownership, 154–6
Self-realisation *see* selves
Self-valuation *see* selves

Sense of justice, 176
Singer, Peter, 37, 59
Sink problems, 188
Sixth extinction, 2
Slavery, 158–9
Smith, Adam, 136, 220
Smuts, J. C., 16
Social animals, 34
Social contract, 107
Social justice, 75, 91, 148–54
Social policy, 83
Socialism, 109, 134, 232
Sociobiology, 34–45
Special obligations, 85–7, 90
Speciesism, 52, 59, 62
Spinoza, 19, 21
Stakeholders, 203, 206
State as 'higher calling', 180
State Investment Agency, 211
States, 120, 136–42, 180, 225
Stewardship, 7, 27–8, 61–2
Strong sustainability, 198
Substance *see* monism
Substitutability of capital, 198
Survival lottery, 160
Sustainable development, 197–200, 203, 213–14
Sylvan, Richard, 44
Systems theory, 16–18

Taylor, Bob Pepperman, 3–5
Taylor, Michael, 141
Third World environmentalism, 196
Toxic waste, 115
Tradition, 172
Tragedy of the commons, 106, 136, 233
Transcendence of morality, 85
Turner, R. K., 197, 198

Universal standpoint, 66
Universalist moral theories, 90, 128–9, 215, 222
Urbanisation, 221
US Supreme Court, 113, 126
Utilitarianism, 146–8

Value
  aesthetic, 68, 78, 146
  individuality, 80–1
  inherent, 56–8, 63
  instrumental, 56
  intrinsic, 19, 55–8, 63–70, 76–80
  intuitionism, 56
  'no value without a valuer', 55–7, 64
  non-instrumental, 51, 57
  objectivism, 58, 64
  scarcity, 77, 80–1
  subjectivism, 65, 67
  urgency, 81
Van Parijs, P., 161n
Variety of species *see* biodiversity
Veil of ignorance, 151

Walzer, Michael, 132
Waste, 189, 192
Weak sustainability, 197–8
Wells, David, 83–9
Western middle class, 193, 196
'Wider' self *see* deep ecology
Williams, Bernard, 84–8
Wilson, Edward, 34, 41–5
Wonderfulness, 67–70, 147
Worker co-operatives, 211–12
World Bank Global Environmental Facility, 214
World Environment Council, 111–12